鲁班工坊数控技术专业系列丛书

数控车床操作与编程实训

吴云飞 主编

天津大学出版社
TIANJIN UNIVERSITY PRESS

图书在版编目（C I P）数据

数控车床操作与编程实训 / 吴云飞主编. -- 天津：
天津大学出版社, 2022.7
　（鲁班工坊数控技术专业系列丛书）
　ISBN 978-7-5618-7194-2

　Ⅰ.①数… Ⅱ.①吴… Ⅲ.①数控机床－车床－操作
－职业教育－教材②数控机床－车床－程序设计－职业教
育－教材 Ⅳ.①TG519.1

中国版本图书馆CIP数据核字(2022)第122876号

组稿编辑　胡小捷
责任编辑　胡小捷
装帧设计　逸　凡

SHUKONG CHECHUANG CAOZUO YU BIANCHENG SHIXUN

出版发行	天津大学出版社
地　　址	天津市卫津路92号天津大学内(邮编:300072)
电　　话	发行部:022-27403647
网　　址	www.tjupress.com.cn
印　　刷	北京盛通商印快线网络科技有限公司
经　　销	全国各地新华书店
开　　本	185mm×260mm
印　　张	15
字　　数	434千
版　　次	2022年7月第1版
印　　次	2022年7月第1次
定　　价	50.00元

前言

为加强我国职业教育与"一带一路"沿线国家职业教育的交流合作，配合马达加斯加"鲁班工坊"项目的实践教学，提高中国职业教育的国际影响力，创新职业院校国际合作模式，输出我国优秀职业教育资源，特编写了本书。

本书面向马达加斯加"鲁班工坊"项目，采用中英文双语编写。本书的编写以"鲁班工坊"数控加工设备为载体，突出"以职业标准为依据，以企业需求为导向，以职业能力为核心"的教学理念，反映新知识、新技术、新工艺、新方法在企业生产实际中的应用，并注重对职业能力的培养。全书共分为 4 个项目 11 个任务，力图做到深入浅出，便于教学。

本书内容编排由浅入深，理实结合，并添加了精选案例，以任务驱动的形式将知识点与技能点有机融合；以科学性、实用性、通用性为原则，使教材符合现代职业教育机械类专业课程体系设置。本书围绕 FANUC 0i TD 数控系统车削加工操作和编程方法，结合实例给出了典型零件完整的加工程序清单及说明。对于编程加工的注意事项、编程技巧等，以"说明""注意"等小栏目进行补充。本书可供各类职业院校数控专业教学使用，也可供数控技术大赛选手参考，还可供使用 FANUC 0i 系统的工程人员和操作人员使用。

本书由天津市机电工艺技师学院吴云飞担任主编，天津职业技术师范大学徐国胜担任主审。参加编写的有天津市机电工艺技师学院许春年、杨福玲、臧成阳三位老师。项目一由许春年编写，项目二由杨福玲编写，项目三由臧成阳编写，项目四由吴云飞编写。

在此特别感谢山东辰榜数控装备有限公司为本书编写提供的帮助和支持，同时本书在编写过程中参阅了许多国内外公开出版与发表的文献，对其作者表示衷心的感谢！限于编者精力及水平有限，书中难免有疏漏，恳请广大读者斧正。

编　者

目录

项目一

数控车床操作基础

【项目描述】

数控机床是以数字运算为核心，运行程序代码，使用计算机统一各部件协调工作来加工零件的冷加工设备，是智能制造的核心设备之一。其主要应用于 IT、汽车、轻工、医疗、航空等行业，是高速率、高精度、高可靠性的机械加工设备。讲解本部分内容应理论联系实际，在介绍基本数控知识的同时，利用多媒体设备进行演示，可以使学生对数控车床、数控系统有更深刻的、直观的了解。

【学习目标】

（1）掌握数控车床安全操作规程，了解数控车床的维护与保养，了解数控车床的应用及各组成部分。

（2）熟悉数控车床相关定义及分类。

（3）掌握 CK6150e 卧式数控车床主要技术规格参数，掌握数控车床的手动操作步骤。

（4）熟悉 FANUC 0i 系统数控车床的面板。

（5）掌握数控车床坐标系与基本点的定义。

（6）掌握各种代码的基本含义。

任务一　　入门教育——认识数控车床

一、任务描述

在生产加工中经常会遇到复杂的轴类零件，为保证加工精度，提高生产效率，一般都选择数控车床进行加工。因此有必要让学生了解这种设备，并且在独立使用数控车床之前，通过阅读资料、观摩操作等方式初步了解生产任务。

二、任务分析

在正式操作数控车床之前，学生应通过阅读机床说明书、教材以及在互联网上下载、阅读相关资料，掌握数控车床安全操作规程，并了解数控车床的维护与保养，为正式操作机床做好准备。

三、相关知识与技能

数控设备是一种自动化程度较高，结构较复杂的先进加工设备，是加工制造企业的重点、关键设备。要发挥数控设备的高效益，就必须正确操作和精心维护，保证设备的利用率。正确地操作使用，能够防止机床非正常磨损，避免突发故障；做好日常维护保养，可使设备保持良好的技术状态，延缓劣化进程，并能及时发现和消灭故障隐患，从而保证安全运行。

1. 数控车床安全操作规程

（1）认真阅读使用说明书及相关操作手册，完全理解后方可操作机床。严禁擅自修改 CNC 机床（数控机床）、PMC（可编程机床控制器）的厂家设置参数，禁止移动或损坏安装在机床上的警告标牌。

（2）开机前应对数控车床进行全面细致的检查，包括操作面板、导轨面、卡爪、尾座、刀架、刀具等，确认无误后方可操作。

（3）数控车床通电后，检查各开关、按钮和按键是否正常、灵活，机床有无异常现象。

（4）机床工作开始前要中、低速空运转 5 min 以上，认真检查润滑系统是否正常工作，如机床已长时间未开动，须先采用手动方式向各部分系统供油润滑。

（5）程序输入后，应仔细核对代码、地址、数值、正负号、小数点及语法是否正确。

（6）正确测量和计算工件坐标系，并对所得结果进行检查。

（7）输入工件坐标系并认真核对坐标、坐标值、正负号、小数点。

（8）未安装工件前，空运行一次程序，检查程序能否顺利运行，刀具和夹具安装是否合理，有无超程现象。使用的刀具必须与机床允许的规格相符合。

（9）检查卡盘夹紧工作的状态，机床开动前，必须关好机床防护门。禁止用手接触刀尖和铁屑，铁屑必须用铁钩或毛刷来清理。禁止用手或其他任何方式接触正在旋转的主轴、工件及机床其他运动部位。

（10）试切削时快速倍率开关必须调到较低挡位。工件伸出车床主轴后端 100 mm 以上时，须在伸出位置设置防护物。

（11）试切削进刀时，在刀具运行至工件 30~50 mm 处，必须在进给保持下，验证 Z 轴和 X 轴坐标剩余值与加工程序是否一致。

（12）试切削和加工中，刃磨刀具和更换刀具后，要重新测量刀具位置并修改刀具补偿值和刀具补偿号。

（13）程序修改后，要对修改部分仔细核对。

（14）操作中出现工件跳动、打抖，异常声音，夹具松动等异常情况时必须停车处理。

（15）紧急停车后，应重新进行机床"回零"操作，才能再次运行程序。

（16）在操作机床时，只允许一个人操作，不得两个人或两个人以上同时操作一台设备。操作者离开机床、变换速度、更换刀具、测量尺寸、调整工件时，都应停车。

（17）在操作结束后，要对机床进行"回零"操作，并认真打扫机床卫生。不允许采用压缩空气清理机床、电气柜及 NC 单元（数字控制单元）。

2. 数控车床的维护与保养

数控车床具有集机、电、液于一体的技术密集和知识密集的特点，是一种自动化程度高、结构复杂且昂贵的先进加工设备。为了充分发挥其作用，减少故障的发生，必须做好日常维护工作，所以要求数控车床维护人员不仅要有机械、加工工艺以及液压、气动方面的知识，也要具备计算机、自动控制、驱动及测量技术等方面的知识，这样才能全面了解、掌握数控车床，及时做好车床维护工作。车床维护工作主要有下列内容。

（1）选择合适的使用环境。数控车床的使用环境（如温度、湿度、振动、电源电压、频率及干扰等）会影响机床的正常运转，故在安装机床时应做到安装条件和要求严格符合机床说明书的规定。在经济条件允许的情况下，应将数控车床与普通机械加工设备分开安装，以便于维修和保养。

（2）应为数控车床配备数控系统编程、操作和维修的专门人员。这些人员应熟悉所用机床的机械、数控系统、强电设备、液压、气压等部分，了解机床使用环境、加工条件等要求，并能按照机床和系统使用说明书的要求正确使用数控车床。

（3）伺服电机的保养。对于数控车床的伺服电机，要 10~12 个月进行一次维护保养，加速或者减速变化频繁的机床要 2 个月进行一次维护保养。维护保养的主要内容有：用干燥的压缩空气吹除电刷的粉尘，检查电刷的磨损情况，如需更换，需选用规格型号相同的电刷，更换后要空载运行一定时间使其与换向器表面吻合；检查清扫电枢整流子以防止短路，如机床安装了测速电机和脉冲编码器，也要对其进行检查和清扫。

（4）及时清扫车床。如及时进行空气过滤网的清扫、电气柜的清扫、印制线路板的清扫等。

（5）机床电缆线的检查。主要检查电缆线的移动接头、拐弯处是否出现接触不良、断线和短路等故障。

（6）及时更换参数存储器电池。有些数控系统的参数存储器是采用 CMOS 元件，其存储内容在断电时靠电池供电保持，一般应在一年内更换一次电池，并且一定要在数控系统通电的状态下进行，否则会使存储参数丢失，导致数控系统不能工作。

（7）长期不用的数控车床的保养。在数控车床闲置不用时，应经常给数控系统通电，在机床锁住的情况下，使其空运行。在空气湿度较大的梅雨季节应该每天通电，利用电器元件运行时本身散发的热量驱走电器柜内的潮气，以保证电子部件的性能稳定可靠。

表 1-1-1 为一台数控车床保养一览表。

表 1-1-1 数控车床保养一览表

序号	检查周期	检查部位	检查要求
1	每天	导轨润滑油箱	检查油量,及时添加润滑油,检查润滑油泵是否能定时启动打油及停止
2	每天	主轴润滑恒温油箱	检查工作是否正常,油量是否充足,温度范围是否合适
3	每天	机床液压系统	检查油箱泵有无异常噪声,工作油面高度是否合适,压力表指示是否正常,管路及各接头有无泄漏
4	每天	压缩空气源压力	检查气动控制系统压力是否在正常范围内
5	每天	X、Z 轴导轨面	清除切屑和脏物,检查导轨面有无划伤损坏,润滑油是否充足
6	每天	各防护装置	检查机床防护罩是否齐全有效
7	每天	电气柜各散热通风装置	检查各电气柜中冷却风扇是否正常工作,风道过滤网有无堵塞,及时清洗过滤器
8	每周	各电气柜过滤网	清洗过滤网上黏附的尘土
9	不定期	冷却液箱	随时检查液面高度,及时添加冷却液,冷却液箱太脏时应及时更换
10	不定期	排屑器	经常清理切屑,检查排屑器有无卡住现象
11	每半年	检查主轴驱动皮带	按说明书要求调整皮带松紧程度
12	每半年	各轴导轨上镶条,压紧滚轮	按说明书要求调整松紧状态
13	每年	检查和更换电机碳刷	检查换向器表面,去除毛刺,吹净碳粉,磨损过多的碳刷应及时更换
14	每年	液压油路	清洗溢流阀、减压阀、滤油器、油箱,过滤液压油或更换液压油
15	每年	主轴润滑恒温油箱	清洗过滤器、油箱,更换润滑油
16	每年	冷却油泵过滤器	清洗冷却油池,更换过滤器
17	每年	滚珠丝杠	清洗丝杠上旧的润滑脂,涂上新润滑脂

四、任务实施

(1)学生通过阅读资料了解、认识、熟悉数控机床安全操作规程以及数控车床的维护与保养。

(2)进入实训车间通过观摩操作等方式巩固已学知识。

五、任务评价

数控车床安全操作及机床保养考核。

(1)考核形式:口试。

（2）考核内容：

①了解数控车床安全操作规程；

②熟悉学生实习守则；

③简述数控车床日常检查项目。

（3）考核要求：以简明扼要的语言回答以上问题，突出要点。

任务二　　数控车床概述及手动操作

一、任务描述

CK6150e 数控车床是目前使用较为广泛的数控机床之一。它主要用于加工轴类零件和盘类零件的内 / 外圆柱面、圆锥面、复杂回转内外曲面以及圆柱、圆锥螺纹等，并能进行切槽、钻孔、扩孔、铰孔等加工操作。CK6150e 数控车床是新一代经济型数控车床，数控装备选用 FANUC 0i TD 系统，其外观如图 1-2-1 所示

图 1-2-1　CK6150e 数控车床外观

CK6150e 数控车床采用卧式机床布局，数控系统控制横（X）纵（Z）两坐标移动，主要承担各种轴类及盘类零件的半精加工及精加工，可加工内 / 外圆柱面、锥面、螺纹、孔以及各种曲线回转体；机床主轴箱采用变频电机实现手动三挡无级调速，刀架

为四刀位；适合教学及企业生产使用。

二、相关知识与技能

1. 数控机床简介

数控机床又称 CNC（Computer Numerical Control）机床，是一种安装了程序控制系统的机床，该系统能逻辑地处理具有控制编码或其他符号指令规定的程序。数字控制是近代发展起来的一种自动控制技术，用数字化的信息对某一对象进行控制，其控制对象为位移、速度、温度、压力、流量、颜色等。

2. 数控车床的组成及分类

1）数控车床的组成

数控车床一般由车床主体、数控装置和伺服系统三大部分组成，各组成部分功能见图 1-2-2 。

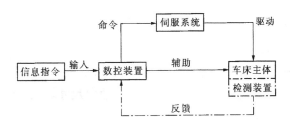

图 1-2-2 数控车床各组成部分功能示意图

（1）车床主体。车床主体是实现加工过程的实际机械部件，主要包括：主运动部件（如卡盘、主轴等）、进给运动部件（如工作台、刀架等）、支承部件（如床身、立柱等）以及冷却、润滑、转位部件和夹紧、换刀机械手等辅助装置。

数控车床主体经过专门设计，各个部位的性能都比普通车床优越，如结构刚性好，能适应高速车削需要；精度高，可靠性好，能适应精密加工和长时间连续工作等。

（2）数控装置和伺服系统。数控车床与普通车床的主要区别为是否具有数控装置和伺服系统这两大部分。如果说，数控车床的检测装置就相当于人的眼睛，那么数控装置相当于人的大脑，伺服系统则相当于人的双手。这样就不难看出这两大部分在数控车床中所处的重要位置了。

①数控装置。数控装置的核心是计算机及运行在其上的软件，它在数控车床中起"指挥"作用，如图 1-2-3 所示。 数控装置接收加工程序送来的各种信息，并经处理和调配后，向驱动机构发出执行命令。 在执行过程中，其驱动、检测等机构同时将有关信息反馈给数控装置，以便经处理后发出新的执行命令。

图 1-2-3　数控装置

②伺服系统。伺服系统通过驱动电路和执行元件（如伺服电机等），准确地执行数控装置发出的命令，完成数控装置所要求的各种位移。

数控车床的进给传动系统常用进给伺服系统来工作，因此也称为进给伺服系统。

进给伺服系统一般由位置控制、速度控制、伺服电动机、检测部件以及机械传动机构五大部分组成。但习惯上所说的进给伺服系统，只是指速度控制、伺服电动机和检测部件三部分，而且将速度控制部分称为伺服单元或驱动器。

2）数控车床的分类

数控车床的种类较多，常用的分类方式有以下几种。

（1）按主轴分布形式分类。

①立式数控车床。立式数控车床的主轴垂直于水平面，并有一个直径较大、用于装夹工件的工作台。立式数控车床主要用于加工径向尺寸较大、轴向尺寸较小的大型复杂零件，如图 1-2-4 所示。

②卧式数控车床。卧式数控车床的主轴平行于水平面，又可分为水平导轨卧式数控车床和倾斜导轨卧式数控车床。图 1-2-5 所示为倾斜导轨卧式数控车床。

图 1-2-4　立式数控车床

图 1-2-5　卧式数控车床

（2）按刀架数量分类。

①单刀架数控车床，如图 1-2-6 所示。

②双刀架数控车床，如图 1-2-7 所示。

图 1-2-6　单刀架数控车床

图 1-2-7　双刀架数控车床

（3）按数控系统功能分类。

①经济型数控车床。经济型数控车床常常基于普通车床数控化改造而成，一般为前置刀架车床，主要用于加工精度要求不高，具有一定复杂形状的零件。图 1-2-1 所示即为经济型数控车床。

②全功能型数控车床。图 1-2-5 所示为全功能型数控车床，这类车床的总体结构先进、控制功能齐全、加工自动化程度较高、辅助功能齐全、稳定可靠性较好，适用于加工精度要求较高、形状复杂的零件。

③数控车铣中心。图 1-2-8 所示的数控车铣中心是以全功能数控车床为主体，并配置刀库、换刀装置、分度装置、铣削动力头和机械手等，能够实现多工序复合加工的机床。在零件一次装夹后，可完成车、铣、钻、扩、铰、攻螺纹等多种加工工序。

图 1-2-8　数控车铣中心

三、任务实施

以 CK6150e 卧式数控车床为例了解车床的技术参数。

1. CK6150e 卧式数控车床主要技术参数

表 1-2-1 为 CK6150e 卧式数控车床主要技术参数。

表 1-2-1 CK6150e 卧式数控车床主要技术参数

	项目	参数值
技术规格	床身最大回转直径 /mm	500
	滑板上最大工件回转直径 /mm	280
	最大加工长度 /mm	1 000
主传动系统（变频无级调速）	主传动形式	变频无级调速
	主轴转速级数	无级
	主轴转速 /（r/min）	200~2 200
	主轴端部结构	C8
	主轴孔直径 /mm	ϕ82
	主轴孔前端锥度	90 公制 1：20
	主电机型号	YVP132M–4–7.5kW
进给系统	刀架最大行程 /mm	X 向：300；Z 向：1 050
	滚珠丝杠直径 × 螺距 /mm	OX 向：ϕ25×5；OZ 向：ϕ40×6
	快速移动进给 /（mm/min）	OX 向：8 000；OZ 向：10 000
	定位精度 /mm	OX 向：0.015；OZ 向：0.020
	重复定位精度 /mm	OX 向：0.010；OZ 向：0.015
	工件加工精度	IT6~IT7
	工件表面结构参数 /μm	Ra1.6
尾座装置	尾座套筒直径 /mm	75
	尾座套筒行程 /mm	170
	尾座套筒锥孔锥度	莫氏 5°
刀架装置	标准配置	电动立式四工位刀架
	特殊选择配置	六（八）工位电动刀架
	重复定位精度 /mm	0.005
	刀杆截面 /mm	25×25
	选择配置	FANUC、西门子、华中、KND
机床外形尺寸及质量	机床外观尺寸（长 × 宽 × 高）/mm	2 800×1 600×1 870
	机床净重 /kg	2 800

2. 手动操作机床

当我们操作机床按照编程程序加工工件时，机床的运行基本上是自动完成的，而不使用编程程序情况下，则要手动操作机床。手动操作机床有如下步骤。

（1）手动返回机床参考点。由于机床采用增量式测量系统，一旦断电后，数控系统就失去了对参考点坐标的记忆，所以当数控系统再次通电后，操作者必须首先进行返回参考点的操作。另外，机床在操作过程中如果遇到急停信号或超程报警信号，待故障排除后，机床恢复工作时，也必须进行返回参考点的操作。具体操作步骤如下：将"MODE"开关置于"ZERO RETUEN"。提醒操作者注意：当滑板上的挡块距离参考点开关不足 30 mm 时，要首先按下"JOG"按钮，使滑板向参考点的负方向移动，直到距离大于 30 mm 时停止移动，然后再返回参考点。分别按下 OX 轴和 OZ 轴的"JOG"按钮，使滑板沿 X 轴或 Z 轴正向移向参考点。在此过程中，操作者应按住"JOG"按钮，直到参考点返回，指示灯亮，再松开按钮。在滑板移动到两轴参考点附近时，会自动减速移动。

（2）滑板的手动进给。当手动调整机床时，或是要求刀具快速移动以接近或离开工件时，需要手动操作滑板进给。滑板进给的手动操作方法有两种：一种是按下"JOG"按钮使滑板快速移动，另一种是摇动手摇轮移动滑板。

（3）快速移动机床。换刀或是手动操作时，要求刀具快速移动以接近或离开工件，其操作步骤如下：将"MODE"开关置于"RAPID"方式；用"APIDOVERRIDE"开关选择滑板快速移动的速度；按下"JOG"按钮，使刀架快速移动到预定位置。

（4）手摇轮进给。手动调整刀具时，要用手摇轮确定刀尖的正确位置，或是试切削时，一边用手摇轮微调进给速度，一边观察切削情况。其操作步骤是：将"MODE"开关转到"HANDLE"位置（可选择 3 个位置），选择手摇轮转动 1 格（将"MODE"开关转至 ×1，手摇轮转 1 格滑板移动 0.001 mm；若指向 ×10，手摇轮转 1 格滑板移动 0.01 mm；若指向 ×100，手摇轮转 1 格滑板移动 0.1 mm）；使手摇轮左侧的 X、Z 轴开关扳向滑板要移动的坐标轴；转动手摇脉冲发生器，使刀架按照指定的方向和速度移动。

（5）主轴的操作。主轴的操作主要包括主轴的启动、停止以及主轴的点动。

①主轴启动与停止。主轴的启动与停止是用来调整刀具或调试机床的。具体操作步骤是：将"MODE"开关置于手动方式（MANU）中任意一个位置；用主轴功能按钮中的"FWD–RVS"开关确定主轴旋转方向，在"FWD"位置，主轴正转，在"RVS"位置，主轴反转；旋转主轴"SPEED"旋钮至低转速区，防止主轴突然加速；按下"START"按钮，主轴旋转，在主轴转动过程中，可以通过"SPEED"旋钮改变主轴的转速，且主轴的实际转速显示在 CRT 显示器上；按下主轴"STOP"按钮，主轴停止转动。

②主轴的点动。当需要使主轴旋转到便于装卸卡爪或是便于检查工件的装夹位置时，需要操作主轴点动。其操作方法是：将"MODE"开关置于自动方式"AUTO"中的任意一个位置；将主轴"FWD–RVS"开关指向所需的旋转方向；按下"START"按钮，主轴转动，按钮抬起，主轴停止转动。

（6）刀架的转位。装卸刀具，测量切削刀具的位置以及对工件进行试切削时，都要在"MDI"状态下编程执行。其操作步骤是：将"MODE"开关置于"MDI"；按下功能键"PRGRM"输入"T10/T20/T30/T40"后再按下"START"按钮。

（7）手动尾座的操作。手动尾座的操作包括尾座体的移动和尾座套筒的移动。

①尾座体的移动。手动操作尾座体使其前进或后退，主要用于轴类零件加工时调整尾座的位置，或是加工短轴和盘类零件时将尾座退至某一合适的位置。其操作步骤

是：将"MODE"开关置于"MANU"方式中的任一位置；按下"TAIL STOCK INTERLOCK"按钮，松开尾座，其按钮上方指示灯亮；移动滑板带动尾座移动至预定位置；再次按下"TAIL STOK INTERLOCK"按钮，尾座被锁紧，指示灯灭。

②尾座套筒的移动。尾座套筒的伸出或退回用于加工轴类零件时，顶尖顶紧或松开工件。其操作方法是：将"MODE"开关置于"MANU"方式中的任一位置，按下"QUILL"按钮，尾座套筒带着顶尖退回，指示灯灭。

（8）卡盘的夹紧与松开。卡盘的夹紧和松开是通过脚踏开关实现的，其操作步骤如下：扳动电箱内卡盘正、反卡开关，选择卡盘正卡或反卡；第一次踏下开关卡盘松开，第二次踏下开关卡盘夹紧。

四、任务评价

数控车床的组成及 CK6150e 卧式数控车床主要技术规格参数考核。

（1）考核形式：口试。

（2）考核内容：

①简述数控车床的组成和分类；

② CK6150e 数控车床主要技术规格参数。

（3）考核要求：以简单扼要的语言叙述以上问题，突出要点。

任务三　　FANUC 系统操作面板

一、任务描述

FANUC 0i 数控系统操作面板由控制器面板（图 1-3-1）和操作面板（图 1-3-2）两大部分组成。

图 1-3-1　FANUC 0i 数控系统控制器面板

图 1-3-2　FANUC 0i 数控系统操作面板

二、任务分析

控制器面板由 CRT 显示部分和 MDI 键盘构成，如图 1-3-1 所示，由 FANUC 系统厂家生产，在 FANUC 系列中控制器面板操作基本相同。对于操作面板，由于生产厂家不同，按键和旋钮的设置有所不同，但功能应用大同小异。对不同厂家生产的数控机床，操作时要灵活掌握按键和旋钮位置。

三、相关知识与技能

1. 控制器面板

控制器面板是人机对话的窗口。如图 1-3-1 所示，控制器面板可显示车床的各种参数和状态，如显示车床参考点坐标、刀具起点坐标、输入数控系统的指令数据、刀具补偿量的数值，显示报警信号、自诊断结果等。在 CRT 显示器的下方有软键操作区，共有 7 个软键，用于各种 CRT 画面的选择。

2.FANUC 0i 操作面板

图 1-3-2 所示操作面板是由机床厂家根据机床功能和结构自行配置的。各厂家配置的操作面板在按键排列和表现形式上各不相同。操作面板一般由监控灯和操作键组成，监控灯用于对机床和数控系统的运行模式进行设置和监控；操作键如急停键、进给倍率旋钮、主轴进行增加或减少按钮、启停键、手摇脉冲发生器等用于实现对机床和数控系统的控制。

（1）系统电源开启及关闭按钮，见表 1-3-1。

表 1-3-1　系统电源开启及关闭按钮

图示	功能说明
	名称：控制器开启按钮（绿色按钮） 操作面板上该按钮触发后，控制器电源开启，数控系统进入启动状态
	名称：控制器关闭按钮（红色按钮） 操作面板上该按钮触发后，控制器电源关闭，数控系统进入关闭状态； 程序运行过程中严禁触发此按钮

（2）紧急停止按钮，见表 1-3-2。

表 1-3-2　紧急停止按钮

图示	功能说明
	名称：紧急停止按钮（红色按钮）。 紧急停止按钮为红色，位于机床操作面板左侧中间位置。 （1）机床在操作或者使用过程中如发生紧急情况，需要立即按下此按钮，按钮按下瞬间可使机床动作全面停止，同时输出到电动机的电流中断，但机床不断电。 （2）按下紧急停止按钮后会出现如下现象： ①如果伺服轴在运行中，则运行的轴停止移动（如机床配有第四轴，并且轴在运转中，第四轴将停止运转）； ②旋转中的主轴停止旋转； ③机床显示器显示报警信息，画面信息为 1000 EMG STOP OR OVERTRAVEL； ④如果刀盘旋转中按下紧急停止按钮，则刀盘立即停止旋转； ⑤如果换刀过程中按下紧急停止按钮，则换刀动作停止，进入换刀异常中断情况。 （3）需要注意以下条件： ①将发生的紧急情况完全解除后，才能解除此按钮； ②停止当时所有指令与机器状况均已被删除时，需要重新核对加工程序，无异常时再进行相关操作； ③当执行自动换刀的中途，按下此按钮所有动作将立即停止，因此刀盘可能位于不确定位置

（3）程序启动和暂停按钮，见表 1-3-3。

表 1-3-3 程序启动和暂停按钮

图示	功能说明
	名称：程序启动按钮 　在自动运转模式（手动输入、记忆、联机）下，选中需要执行的加工程序，按下"程序启动"按钮后，程序即开始执行
	名称：程序暂停按钮 　（1）在自动运转模式（手动输入、记忆、联机）下，按下"程序暂停"按钮后，各轴立即减速停止，进入运转休止状态； 　（2）当再按下"程序启动"按钮后，加工程序将从当前暂停的单节继续执行

（4）程序保护开关，见表 1-3-4。

表 1-3-4 程序保护开关

图示	功能说明
	名称：程序保护开关 （1）为防止本机床控制器中的程序被他人编辑、取消、修改、建立，钥匙应交由专人保管； （2）一般情况，将此钥匙设定在"OFF"的位置，以确保程序不被修改或删除； （3）如果想对程序予以编辑、取消或修改时，应将本钥匙设定在"ON"的位置

（5）操作模式按钮，见表 1-3-5。

表 1-3-5 操作模式按钮

图示	功能说明
	名称：编辑模式 机床显示器左下角显示"编辑"。 （1）可以对程序予以编辑、修改、增加或删除，需将程序保护开关钥匙设定在"ON"状态，在此模式下才可编辑； （2）该模式仅用于编辑，不能执行程序； （3）执行新编辑程序功能时，必须转至自动模式（手动输入、记忆），才会执行编制程序； （4）程序编辑完成，控制器即会自动存储，不必再执行存储动作； （5）该模式下，可由各计算机读入加工程序
	名称：自动模式 机床显示器左下角显示"MEM"。 （1）此模式下，按下"程序启动"按钮，即可执行当前已经选择的加工程序； （2）此模式下，能执行 CNC 内存中的程序； （3）此模式下的进给速率，参照进给倍率调整按键说明； （4）此模式下，程序在执行 M30 时即为程序结束，同时使程序还原
	名称：MDI 手动输入模式 机床显示器左下角显示"MDI"。 （1）此模式下，在控制器 MDI 面板输入单节程序指令予以执行； （2）此模式下，指令执行完成后，可通过设置参数来确定编制的程序是否需要消除； （3）此模式下，仅能输入部分程序段

续表

图示	功能说明
	名称：手摇模式 机床显示器左下角显示"HAND"。 （1）此模式下，可用手持单元移动各轴； （2）此模式下，轴向移动可由手持单元上的轴向旋钮选择轴向，从而控制选中轴向的移动距离； （3）此模式下，各轴手动移动速度可由手持单元上的进给倍率旋钮确定； （4）手摇脉冲发生器转动速度不能大于 5 圈 /s
	名称：手动模式 机床显示器左下角显示"JOG"。 （1）此模式下，欲移动各轴，可按各轴向键并选择慢速进给率； （2）此模式下，移动进给速率按照慢速执行，速率调整范围为 0~1 000 mm/min； （3）此模式下，按轴向键时，其指定轴即可移动，松开按键，轴即停止； （4）此模式下，配合快速按钮使用时，移动进给速率按照快速执行，按动各轴向键，依指定轴向移动，松开按键，轴即停止
	名称：原点复归模式 机床显示器左下角显示"REF"。 （1）在手动方式下用于伺服轴回原点操作； （2）机床每次开机后，需做一次原点复归操作。如果各轴位置在原点附近，则需要手动移动各轴远离原点位置后，再继续完成原点复归操作； （3）在该模式下，选择需机械原点复归的轴向键后，原点指示灯持续闪烁，直到原点复归动作完成后，原点指示灯不再闪烁，处于常亮状态； （4）各轴的机械原点复归速率为：参数设定回零速度 × 进给倍率开关所在的倍率值（%）

（6）辅助功能开关，见表 1-3-6。

表 1-3-6　辅助功能开关

图示	功能说明
	名称：单节执行开关 此功能仅在自动相关模式下有效。 （1）该按键提示灯亮时，程序单节执行功能键有效。此功能打开后，程序将按单节执行，执行完当前单节后程序暂停，继续按程序启动按钮后方可执行下一单节程序，以后执行程序依此类推； （2）当按键指示灯不亮时，程序单节执行功能键无效，加工程序将一直被执行到程序终了
	名称：空运行开关 此功能仅在自动相关模式下有效。 （1）此按键指示灯亮时，OZ 轴锁定功能键有效，此功能打开后，程序中所设定的 F 值（切削进给率）指令无效，其各轴移动速率依慢速位移速率移动； （2）在功能无效时，若程序执行循环指令，慢速进给率或切削进给率无法改变进给率，按照控制中的 F 值以固定的进给速率运动
	名称：选择跳过开关 此功能仅在自动相关模式下有效。 （1）此按键指示灯亮时，程序选择跳过功能键有效，此功能打开后，自动运行中，当在程序段的开头指定了一个"/"（斜线）符号时，此程序段将略过不被执行； （2）当按键指示灯不亮时，程序选择跳过功能键无效，此功能关闭后，即程序单节前有"/"（斜线）符号时，此程序段也可以正常执行

图示	功能说明
	名称：选择停开关 此功能仅在自动模式下有效。 （1）本按键指示灯亮时，程序选停功能键有效，此功能打开后，执行程序中，若有 M01 指令时，程序将停止于该单节，若需继续执行程序，按下程序启动键即可； （2）当按键指示灯不亮时，程序选停功能键无效，此功能关闭后，即使程序中有 M01 指令，程序也不会停止执行
	名称：机械锁定开关 （1）此按键指示灯亮时，所有轴机械锁定功能键有效。此功能打开后，无论在手动模式或自动模式中移动任意一个轴，CNC 均停止向该轴伺服电机输出脉冲（移动指令），但依然在进行指令分配，对应轴的绝对坐标和相对坐标也得到更新；M、S、T、B 码会继续执行，不受机械锁定限制。 （2）解除该功能后需要重新回归机械零点，在回零正确且完毕后，再进行其他相关操作。如果未回零而进行了相关操作则会造成坐标偏移，甚至出现撞机、程序乱跑等异常现象，从而导致故障或危险发生
	名称：F1 键 此按键依据机床实际配置。预备空格键，操作人员不能进行操作
	名称：F2 键 此按键依据机床实际配置。预备空格键，操作人员不能进行操作
	名称：F3 键 此按键为工作灯扩展键，控制工作灯开启和关闭，不受任何操作模式限制

（7）主轴功能开关，见表 1-3-7。

表 1-3-7　主轴功能开关

图示	功能说明
 主轴降速 ⊖	名称：主轴速度降低开关 （1）此按键位于机床操作面板上，用于降低编程制定的主轴转速 S，实际转速 = 编程给定 S 指令值 × 主轴速度降低倍率值； （2）此开关与设定的主轴转速配合使用
 主轴升速 ⊕	名称：主轴速度提高开关 （1）此按键位于机床操作面板上，用于提高编程制定的主轴转速 S，实际转速 = 编程给定 S 指令值 × 主轴速度提高倍率值； （2）编程设定速度超过主轴最高转速，转速达到 100% 以上倍率时，主轴修调速度等于主轴最高转速； （3）此开关与设定的主轴转速配合使用
 主轴正转 ↻	名称：主轴正转开关 （1）在机床执行一次 S 代码后，选中手动操作模式，按主轴正转按键后，主轴进行顺时针旋转。主轴旋转速度 = 先前执行的主轴速度 S 值 × 主轴修调旋钮所在的挡位。 （2）使用条件： ①仅在手动、快速、手摇模式下才能使用； ②在自动模式下，当程序执行主轴正转 M03 指令后，此按键指示灯会亮。 （3）"主轴停止"或"主轴反转"生效时，指示灯即熄灭。 （4）需要进行主轴反向旋转时，必须使主轴停止后才可指定反向旋转操作
 主轴停止 ○	名称：主轴停止开关 （1）主轴无论处于正转或反转状态下，按此键均可以停止正在旋转中的主轴。 （2）使用条件： ①此按键仅在手动、快速、手摇下才能使用； ②在自动操作时无效。 （3）主轴停止时此按键指示灯会亮，但如果"主轴正转"或"主轴反转"生效时，指示灯即熄灭
 主轴反转 ↺	名称：主轴反转开关 （1）在本机床执行一次 S 代码后，选中手动操作模式，按主轴反转按键后，主轴进行逆时针旋转。主轴旋转速度 = 先前执行的主轴速度 S 值 × 主轴修调旋钮所在的挡位。 （2）使用条件： ①仅在手动、快速、手摇模式下才能使用； ②在自动模式下，当程序中执行主轴反转指令 M04 后，本按键指示灯会亮。 （3）主轴反转时本按键指示灯会亮，但如果"主轴正转"或"主轴停止"生效时，指示灯即熄灭。 （4）需要进行主轴正向旋转时，必须使主轴停止后才可指定正向旋转操作

（8）手动、快速、手摇模式下辅助功能开关，见表 1-3-8。

表 1-3-8　手动、快速、手摇模式下辅助功能开关

图示	功能说明
 冷却	名称：冷却开关 （1）在手动、快速、手摇模式下，按此键指示灯亮后，冷却液喷出； （2）按"RESET"键，冷却液停止喷出，冷却停止，指示灯灭； （3）冷却液开启时需注意冷却液喷嘴的朝向

图示	功能说明
	名称：手动换刀开关 在手动、快速、手摇模式下，每按此键一次，刀具按加工方向旋转一个刀位

（9）轴向选择开关，见表1-3-9。

表1-3-9　轴向选择开关

图示	功能说明
	名称：+X控制按键 在JOG模式下，按住此键则X轴依进给倍率或快速倍率的速度向机床X轴"+"方向（正方向）移动，同时按键指示灯亮；当松开按键后，轴停止向"+"方向移动，同时按键指示灯熄灭。令外，此按键也作为X轴回零触发键
	名称：−X控制按键 在JOG模式下，按住此键则X轴依进给倍率或快速倍率的速度向机床X轴"−"方向（负方向）移动，同时按键指示灯亮；当松开按键后，轴停止向"−"方向移动，同时按键指示灯熄灭
	名称：+Z控制按键 在JOG模式下，按住此键则Z轴依进给倍率或快速倍率的速度向机床Z轴"+"方向（正方向）移动，同时按键指示灯亮；当松开按键后，轴停止向"+"方向移动，同时按键指示灯熄灭。令外，此按键也作为Z轴回零触发键
	名称：−Z控制按键 在JOG模式下，按住此键则Z轴依进给倍率或快速倍率的速度向机床Z轴"−"方向（负方向）移动，同时按键指示灯亮；当松开按键后，轴停止向"−"方向移动，同时按键指示灯熄灭。 另外，当执行Z轴负方向移动程序指令时，此键指示灯也将点亮，停止移动指令时，该按键指示灯熄灭
	名称：超程释放开关 （1）当机床的各轴行程超过硬限位时，机床会出现超程报警，机床的动作停止，这时按住此键，在手轮模式下用手持单元将机床超程的轴反方向移动； （2）绝对式编码器机床超程无需按此键
	名称：手动快移开关 本功能仅在手动模式下有效。手动模式下按下此键指示灯点亮。实际快速进给速度＝参数设置G00最大速度值 × 快速倍率开关所在的倍率值（%）

（10）进给倍率及进给修调开关，见表1-3-10。

表 1-3-10　进给倍率及进给修调开关

表 1-3-10　进给倍率及进给修调开关

图示	功能说明
	名称：进给倍率及进给修调开关 （1）此旋钮位于机床操作面板上，控制编程指定 G01 速度，实际进给速度 = 编程给定 F 指令值 × 进给倍率开关所在倍率值（%）； （2）手动模式下，此时控制 JOG 进给倍率，实际 JOG 进给速度 = 参数设定固定值 × 进给倍率开关所在倍率值（%）； （3）此开关与设定的轴进给速度配合使用

（11）快速倍率开关，见表 1-3-11。

表 1-3-11　快速倍率开关

图示	功能说明
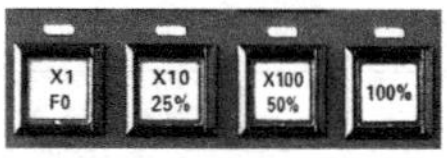	名称：快速倍率开关 （1）此按钮位于机床操作面板上，控制编程指定 G00 速度，实际进给速度 = 参数设置 G00 最大速度值 × 快速进给倍率按钮所在倍率值（%）； （2）快速模式下，此时控制手动快速进给倍率，实际快速进给速度 = 参数设置 G00 最大速度值 × 快速进给倍率按钮所在倍率值（%），快速移动倍率可以在 F0、25%、50%、100% 四个挡位调整； （3）此开关与设定的轴进给速度配合使用

（12）手摇脉冲开关，见表 1-3-12。

表 1-3-12　手摇脉冲开关

图示	功能说明
	名称：轴向选择开关 （1）此开关位于手持单元上，与手轮进给倍率"×1""×10""×100"互相配合使用； （2）此开关用于手轮模式； （3）此开头拨到"0"位不选择任何轴，拨到"X"位选择 OX 轴，拨到"Y"位选择 OY 轴，拨到"Z"位选择 OZ 轴，拨到"4"位选择 4 轴
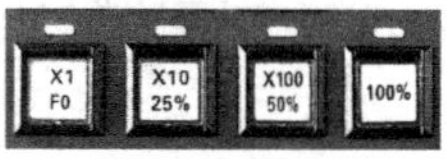	名称：倍率选择开关 （1）此按钮位于手持单元上，与手轮进给倍率"×1""×10""×100"互相配合使用； （2）此按钮用于手轮模式； （3）此按钮按到"×1"位选择手轮进给倍率为 0.001 mm/ 格，按到"×10"选择手轮进给倍率为 0.01 mm/ 格，按到"×100"选择手轮进给倍率为 0.1 mm/ 格
	名称：手轮 MPG （1）本码盘仅在"手轮"模式下有效，用于操作进给轴方向与速度； （2）手摇脉冲发生器的回转方向以顺时针方向为正（即正向旋转后伺服轴往正方向移动），以逆时针方向为负（即负向旋转后伺服轴往负方向移动）。 注意： （1）手轮旋转速度不得大于 5 圈 /s。如果手轮旋转速度超过了 5 圈 /s，刀具有可能在手轮停止旋转后还不能停止下来或者刀具移动的距离与手轮旋转的刻度不符； （2）手持单元使用时应轻拿轻放、注意保护

（13）轴选择、倍率选择与手轮每格的伺服轴移动量关系，见表 1–3–13。

表 1-3-13　轴选择、倍率选择与手轮每格的伺服轴移动量关系

倍率选择	×1	×10	×100
公制移动量	0.001 mm/格	0.01 mm/格	0.1 mm/格
英制移动量	0.000 1 in/格	0.001 in/格	0.01 in/格

注：1 in=25.4 mm。

四、任务实施

1. 编辑程序

将操作面板中"MODE SELECT"旋钮 切换到"EDIT"挡，在 MDI 键盘上按 键，进入编辑页面，选定了一个数控程序后，此程序显示在 CRT 界面上，可对数控程序进行编辑操作。

（1）移动光标。按 PAGE 或 翻页，按 CURSOR 或 移动光标。

（2）插入字符。先将光标移到所需位置，点击 MDI 键盘上的数字或字母键，将代码输入输入域中，按 键，把输入域的内容插入光标所在代码后面。

（3）删除输入域中的数据。按 键删除输入域中的数据。

（4）删除字符。先将光标移到所需删除字符的位置，按 键删除光标所在处的字符。

（5）查找。输入需要搜索的字母或代码，按 CURSOR 开始在当前数控程序中光标所在位置后搜索（代码可以是一个字母或一个完整的代码，例如"N0010""M"等）。如果此数控程序中有所搜索的代码，则光标停留在找到的代码处；如果此数控程序中光标所在位置后没有所搜索的代码，则光标停留在原处。

（6）替换。先将光标移到需要替换字符的位置，将替换成的字符通过 MDI 键盘输入输入域中，按 键，用输入域的内容替代光标所在的代码。

2. 显示数控程序目录

将操作面板中"MODE SELECT"旋钮 切换到"EDIT"挡，在 MDI 键盘上按 键，进入编辑页面，再按软键 ，数控程序名显示在 CRT 界面上。

选择一个数控程序　将操作面板中"MODE SELECT"旋钮 切换到"EDIT"或"AUTO"挡，在 MDI 键盘上按 键，进入编辑页面，按 键输入字母"O"；按数字键输入搜索的号码××××（搜索号码为数控程序目录中显示的程序号），按 CURSOR 开始搜索。找到后，"O××××"显示在屏幕右上角程序号位置，NC 程序显示在屏幕上。

3. 删除一个数控程序

将操作面板中"MODE SELECT"旋钮 [图] 切换到"EDIT"挡，在 MDI 键盘上按 [PRGRM] 键，进入编辑页面，按 [O] 键输入字母"O"；按数字键输入要删除的程序的号码××××；按 [DELET] 键，程序即被删除。

新建一个 NC 程序 将操作面板中"MODE SELECT"旋钮 [图] 切换到"EDIT"挡，在 MDI 键盘上按 [PRGRM] 键，进入编辑页面，按 [O] 键输入字母"O"；按数字键输入程序号，但不可以与已有的程序号重复；按 [INSRT] 键，开始输入程序；每输入一个代码，按 [PRGRM] 键，输入域中的内容显示在 CRT 界面上，用回车换行键 [EOB] 结束一行的输入后换行。

注：MDI 键盘上的数字/字母键，第一次按下时输入的是字母，以后再按下时均为数字。若要再次输入字母，须先将输入域中已有的内容显示在 CRT 界面上（按 [INSRT] 键，可将输入域中的内容显示在 CRT 界面上）。

4. 删除全部数控程序

将操作面板中"MODE SELECT"旋钮 [图] 切换到"EDIT"挡，在 MDI 键盘上按 [PRGRM] 键，进入编辑页面，按 [O] 键输入字母"O"；按 [M] 键输入"−"；按 [O] 键输入"9999"；按 [DELET] 键，全部数控程序即被删除。

5. 程序模拟练习

NC 程序输入后，可检查运行轨迹。首先将光标移到程序头位置，同时按下 [MLK] [DRN] [AUX LOCK] 这三个按键，再将操作面板中"MODE SELECT"旋钮 [图] 切换到"AUTO"挡，按控制面板上 [AUX GRAPH] 按钮，转入检查运行轨迹模式；再按操作面板上的 [START] 按钮，即可观察数控程序的运行轨迹。通过程序模拟，可以校验输入的程序正确与否。程序模拟时，暂停运行、停止运行、单段执行等功能键同样有效。

五、任务评价

数控车床各按键功能及程序输入、校验考核。

（1）考核形式：口试、实际操作。

（2）考核内容：

①熟悉数控车床各按键功能；

②输入程序；

③校验程序。

（3）考核要求：能以简单扼要的语言回答口试问题，突出要点；能独立完成相关实际操作。

任务四　基本编程指令与坐标系的确定

一、任务描述

在数控机床上加工零件并保证零件的加工精度，其实质是保证工件和刀具的相对运动精确无误。所以想要在编程时控制工件和刀具运动，首先需要掌握数控机床上常用的两个坐标系——机床坐标系和工件坐标系；其次要掌握机床参考点、换刀点等内容；最后需根据零件图纸的精度和技术要求等，分析确定零件的工艺过程、工艺参数等内容，用规定的数控编程代码和程序格式编制出合适的数控加工程序。

二、任务分析

为了简化编程和保证程序的通用性，国际化标准组织已经对数控机床的坐标系和方向、命名制定了统一的标准。

同时，数控系统的种类繁多，为实现系统兼容，国际标准化组织制定了相应的标准，我国也在国际标准基础上制定了标准。数控技术的高速发展和市场竞争等因素，导致不同系统间存在部分不兼容，如 FANUC 0i 系统编制的程序无法在 SIEMENS 系统上运行。因此编程必须注意具体的数控系统或机床，应该严格按机床编程手册中的规定进行程序编制。但从数控加工内容本质上讲，各数控系统的各项指令都是应实际加工工艺要求而设定的。

三、相关知识与技能

1. 坐标系的确定

数控车床有三个坐标系，即机械坐标系、编程坐标系和工件坐标系。机械坐标系的原点是生产厂家在制造机床时的固定坐标系原点，也称机械零点。它是在机床装配、调试时已经确定下来的，是机床加工的基准点。在使用中机械坐标系是由参考点来确定的，机床系统启动后，进行返回参考点操作，机械坐标系就建立了。坐标系一经建立，只要不切断电源，坐标系就不会变化。编程坐标系是编制程序时使用的坐标系，一般把 Z 轴与工件轴线重合，X 轴放在工件端面上。工件坐标系是机床进行加工时使用的坐标系，它应该与编程坐标系一致。能否让编程坐标系与工件坐标系一致，是操作的关键。

数控机床上的坐标系采用右手直角笛卡儿坐标系，如图 1-4-1 所示。右手的大拇指、食指和中指保持相互垂直，拇指的方向为 X 轴的正方向，食指为 Y 轴的正方向，中指为 Z 轴的正方向。

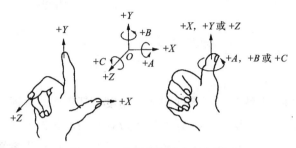

图 1-4-1　右手直角笛卡儿坐标系

2. 运动方向的确定

机床某一部件运动的正方向，是增大工件和刀具之间距离的方向，如图 1-4-2 所示。

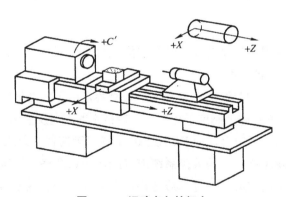

图 1-4-2　运动方向的规定

（1）Z 轴与主轴轴线重合。设 Z 轴远离工件、向尾座移动刀具的方向为正方向（即增大工件和刀具之间距离），向卡盘移动方向为负方向。

（2）OX 轴垂直于 Z 轴。OX 坐标的正方向是刀具离开旋转中心线的方向，反之为负。

3. 数控车床的有关点

1）机床原点

如图 1-4-3 所示，数控车床的坐标系规定，通常把传递切削力的主轴定为 Z 轴。数控车床的机床原点一般设在主轴回转中心与卡盘后端面的交线上，如图 1-4-4 中

的 O 点。

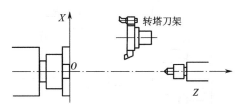

图 1-4-3　卧式刀塔 CNC 车床坐标系

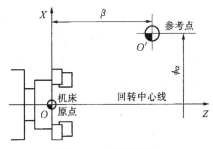

图 1-4-4　机床原点

2）机床参考点

机床参考点也是机床上一个固定的点，它是用机械挡块或电气装置来限制刀架移动的极限位置，作用主要是给机床坐标系一个定位。因为每次开机后无论刀架停留在哪个位置，系统都把当前位置设定为（0,0），这样势必造成基准的不统一，所以每次开机的第一步操作为参考点回归（有的称为回零点），也就是通过确定参考点来确定机床坐标系的原点（0,0）。参考点返回就是使刀架按指令自动地返回到机床的这一固定点，此功能也用来在加工过程中检查坐标系的正确与否和建立机床坐标系，以确保精确地控制加工尺寸。这个点常用来作为刀具交换的点，如图 1-4-4 中的 O 点，$\phi\alpha$、β，为机床 X 轴、Z 轴方向极限行程距离，即机床的理论加工范围。

当机床刀架返回参考点之后，刀架基准点在该机床坐标系中的坐标值即为一组确定的数值。机床在通电之后，返回参考点之前，不论刀架处于什么位置，CRT 上显示的 Z 与 X 坐标值均为 0，只有完成返回参考点操作后，CRT 上的值才立即显示出刀架基准点在机床坐标系中的坐标值，即建立了机床坐标系。

3）工件坐标系原点

在编制程序时，首先要根据被加工零件的形状特点和尺寸，将零件图上的某一点设定为编程坐标原点，该点称为工件坐标系原点，如图 1-4-5 中的 O 点。

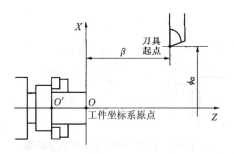

图 1-4-5　工件坐标系

只有使零件上的所有几何元素都有了确定的位置，才能进行路线安排、数值处理和程序编制等，同时也决定了在数控加工时零件在机床上的安放方向。从理论上讲，工件坐标系的原点选在工件上任何一点都可以，但这可能带来烦琐的计算问题，增添编程的困难。为了计算方便，简化编程，通常是把工件坐标系的原点选在工件的回转中心上，具体位置可考虑设置在工件的左端面（或右端面）上，尽量使编程基准与设计基准、定位基准重合。

（1）把坐标系原点设在卡盘面上，如图 1-4-6 所示。

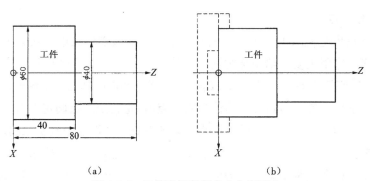

（a）　　　　　　　　　　　　　　（b）

图 1-4-6　工件坐标系原点在卡盘面上

（a）加工图纸上的坐标和尺寸；（b）车床上 CNC 指令的坐标系（与编程坐标系重合）

（2）把坐标系原点设在零件端面上，如图 1-4-7 所示。

在数控机床上加工零件时，刀具与工件的相对运动必须在确定的坐标系中进行，编程人员必须熟悉机床坐标系。规定数控机床的坐标轴及运动方向，是为了准确地描述机床的运动，简化程序的编制方法，并使程序具有互换性。

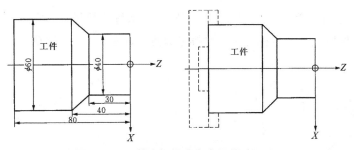

图 1-4-7　工件坐标系原点在零件端面上

机床坐标系是机床唯一的基准，所以必须弄清楚程序原点在机床坐标系中的位置，通常这一步在接下来的对刀过程中完成。对刀的实质是确定工件坐标系的原点在唯一的机床坐标系中的位置。对刀是数控加工中的主要操作和重要技能，对刀的准确性决定了零件的加工精度，同时，对刀效率直接影响数控加工效率。

4）换刀点

当数控车床在加工过程中需要换刀时，在编程时应考虑选择合适的换刀点。所谓换刀点是指刀架转位换刀的位置。数控车床上确定了工件坐标后，换刀点可以是某一固定点，也可以是相对于工件原点位置的任意一点。换刀点应设在工件或夹具的外部，以刀架转位换刀时不碰工件及其他部位为准。

4. 编程的方式

（1）绝对值编程：根据预先设定的编程原点，计算出绝对值坐标尺寸进行编程的一种方法。FANUC 系统中用 X、Z 表示绝对值坐标，其中 X 表示直径值。

（2）增量值编程：根据与前一位置的坐标值增量来表示位置的一种编程方式。用 U、W 表示增量值坐标，其中 U 表示直径增量。

（3）混合编程：绝对值和增量值编程混合起来进行编程的方法。

5. FANUC 数控系统的功能指令

数控机床在编程时，对加工过程中的各个动作，如机床主轴的开、停、换向，刀具的进给方向，冷却液的开、关等，都要用指令的形式给予规定，这类指令称为功能指令。数控程序所用的功能指令，主要有准备功能 G 指令、辅助功能 M 指令、进给功能 F 指令、主轴转速功能 S 指令和刀具功能 T 指令等几种。在数控编程中，用各种 G 指令和 M 指令来描述工艺过程和运动特征。现国际上广泛采用 ISO—1056—l975E 标准，我国等效采用该标准制定了 GB/T 38267—2019 标准。

1）FANUC 数控系统准备功能指令（G 指令）

准备功能指令又称 G 指令或 G 代码，它是建立机床或控制数控系统工作方式的一种指令。这种指令在数控装置插补运算之前需预先规定，为插补运算、刀具补偿运算、固定循环等做好准备。G 指令由字母 G 和其后两位数字组成。表 1-4-1 为 FANUC 系统数控车床常用 G 指令的列表。

表 1-4-1　FANUC 系统数控车床常用 G 指令

G 代码	组	功能	G 代码	组	功能
*G00	01	定位（快速移动）	G55	14	选择工件坐标系 2
G01		直线切削	G56		选择工件坐标系 3
G02		圆弧插补（CW，顺时针）	G57		选择工件坐标系 4
G03		圆弧插补（CCW，逆时针）	G58		选择工件坐标系 5
G04	00	暂停	G59		选择工件坐标系 6
G20	06	英制输入	G70	00	精加工循环
G21		公制输入	G71		内、外径粗切削循环

续表

G 代码	组	功能	G 代码	组	功能
G27	00	检查参考点返回	G72	00	台阶粗切削循环
G28		参考点返回	G73		成形重复循环
G29		参考点返回	G74		Z 向进给钻削
G30		回到第二参考点	G75		X 向切槽
G32	01	切螺纹	G76		切螺纹循环
*G40	07	取消刀尖半径偏置	G90	01	（内、外直径）切削循环
G41		刀尖半径偏置（左侧）	G92		切螺纹循环
G42		刀尖半径偏置（右侧）	G94		（台阶）切削循环
G50	00	主轴最高转速设置（坐标系设定）	G96	02	恒线速度控制
G52		设置局部坐标系	*G97		恒线速度控制取消
G53		选择机床坐标系	G98	05	指定每分钟移动量
*G54	14	选择工件坐标系 1	*G99		指定每转移动量

注：1. 有标记"*"的指令为开机时即已被设定的指令；

2. "00"组别的 G 指令属非模态指令，它们的指令只能在一个程序段中有作用；

3. 一个程序段中可使用若干个不同组群的 G 指令，若使用一个以上同组群的 G 指令则最后一个 G 指令有效。

G 指令从功能上可分为三种。

（1）加工方式 G 代码，执行此类 G 代码时机床有相应动作。在编程格式上必须指定相应坐标值，如"G01 X60. Z0；"。

（2）功能选择 G 代码，相当于功能开与关的选择，编程时不用指定地址符。数控机床通电后具有的内部默认功能一般有设定绝对坐标方式编程、使用米制长度单位量纲、取消刀具补偿、主轴和切削液泵停止工作等状态作为数控机床的初始状态。

（3）参数设定或调用 G 代码，如 G50 坐标设定指令，执行时只改变系统坐标参数；如 G54 执行时只调用系统参数，机床不会产生动作。

2）FANUC 数控系统辅助功能指令（M 指令）

辅助功能指令又称 M 指令或 M 代码。这类指令的作用是控制机床或系统的辅助功能动作，如冷却泵的开、关，主轴的正转、反转，程序结束等。在同一程序段中，若有两个或两个以上辅助功能指令，则读后面的指令。M 指令由字母 M 和其后两位数组成。 FANUC 系统数控车床常用的 M 指令见表 1-4-2。

表 1-4-2　FANUC 系统数控车床常用的 M 指令

M 功能	含义	M 功能	含义
M00	程序停止	M08	开始注入切削液
M01	计划停止	M09	停止注入切削液
M02	程序结束	M30	程序结束并返回开始处
M03	主轴顺时针旋转	M98	调用子程序

M 功能	含义	M 功能	含义
M04	主轴逆时针旋转	M99	子程序返回
M05	主轴停止旋转		

3）FANUC 系统的其他常用功能

一个标准的程序除了必须应用 G 指令和 M 指令外，编程时还应有 F 功能、S 功能和 T 功能。

（1）F 功能也称进给功能，其作用是指定刀具的进给速度。程序中用 F 和其后面的数字组成，F 码可用每分钟进给 G98 指令和每转进给 G99 指令来设定进给单位。

（2）S 功能也称主轴转速功能，其作用是指定主轴的转动速度，程序中用 S 和其后的数字组成。

（3）T 功能也称为刀具功能，其作用是指定刀具号码和刀具补偿号码。程序中用 T 和其后的数字表示，依据机床装刀数的不同可采用二位或四位数字。

6. FANUC 系统数控编程的格式

1）数控程序编制的基本方法

（1）数控程序编制的内容及步骤。如图 1-4-8 所示，编程工作主要包括以下内容。

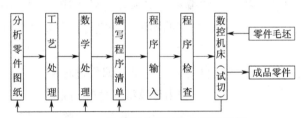

图 1-4-8　数控程序编制的内容及步骤

①分析零件图样和制定工艺方案。对零件图样进行分析，明确加工内容和要求；确定加工方案；选择适合的数控机床；选择或设计刀具和夹具；确定合理的加工路线及切削用量等。这一工作要求编程人员能够按照图样对零件的技术特性、几何形状、尺寸及工艺要求进行分析，并结合数控机床使用的基础知识，如数控机床的规格、性能、数控系统的功能等，确定加工方法和加工路线。

②数值计算。在制定加工工艺方案后，就需要根据零件的几何尺寸、加工路线等，计算刀具中心运动轨迹，以获得刀位数据。数控系统一般均具有直线插补与圆弧插补功能，对于加工由圆弧和直线组成的较简单的平面零件，只需要计算出零件轮廓上相邻几何元素交点或切点的坐标值，得出各几何元素的起点、终点、圆弧的圆心坐标值等，就能满足编程要求。当零件的几何形状与控制系统的插补功能不一致时，就需要进行较复杂的数值计算，一般需要使用计算机辅助计算，否则难以完成。

③编写零件加工程序。在完成上述工艺处理及数值计算工作后，即可编写零件加工程序。程序编制人员使用数控系统的程序指令，按照规定的程序格式，逐段编写加

工程序。程序编制人员只有对数控机床的功能、程序指令及代码十分熟悉，才能编写出正确的加工程序。

④检验程序和首件试切。将编写好的加工程序输入数控系统，就可控制数控机床的加工工作。一般在正式加工之前，要对程序进行检验。通常可采用机床空运转的方式，来检查机床动作和运动轨迹正确性，以检验程序。在具有图形模拟显示功能的数控机床上，可通过显示走刀轨迹或模拟刀具对工件的切削过程，对程序进行检查。对于形状复杂和加工要求高的零件，也可采用铝件、塑料或石蜡等易切材料进行试切削来检验程序。通过检查试件，不仅可确认程序是否正确，还可知道加工精度是否符合要求。若能采用与被加工零件材料相同的材料进行试切，则更能反映实际加工效果，当发现加工的零件不符合加工技术要求时，可修改程序或采取尺寸补偿等措施。

（2）数控加工程序编制的方法。数控加工程序的编制方法主要有两种：手工编程和计算机自动编程。

①手工编程。一般对几何形状不太复杂的零件，所需的加工程序不长，计算比较简单，用手工编程比较合适。

手工编程的特点：手工编程不需要计算机、编程器、编程软件等辅助设备，只需要合格的编程人员，具有编程速度快、及时的优点。但缺点是耗费时间较长，容易出现错误，无法胜任复杂形状零件的编程。据国外资料统计，当采用手工编程时，一段程序的编写时间与其在机床上运行加工的实际时间之比，平均约为 30：1，而数控机床不能开动的原因中有 20%~30% 是由于加工程序编制困难，编程时间较长。

②计算机自动编程。计算机自动编程是指在编程过程中，除了分析零件图样和制定工艺方案由人工进行外，其余工作均由计算机软件辅助完成。

采用计算机自动编程时，数学处理、编写程序、检验程序等工作是由计算机自动完成的，由于计算机可自动绘制出刀具中心运动轨迹，使编程人员可及时检查程序是否正确，需要时可及时修改，以获得正确的程序。又由于计算机自动编程代替程序编制人员完成了烦琐的数值计算，可提高编程效率几十倍乃至上百倍，因此解决了手工编程无法解决的许多复杂零件的编程难题。因而，自动编程的特点就在于编程工作效率高，可解决复杂形状零件的加工编程难题。

2）FANUC 数控系统数控编程的格式

（1）程序的格式。编写加工程序就是按机床动作和刀具路线的实际顺序书写控制指令。把按顺序排列的各指令称为程序段。为了进行连续加工，需要很多程序段，这些程序段的集合称为程序。为识别各程序段所加的编号称为顺序号，而为识别各个程序所加的编号称为程序号。一个完整的程序，一般由程序号、程序内容和程序结束三部分组成。其格式如下：

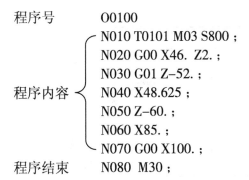

<pre>
程序号   O0100
            N010 T0101 M03 S800；
            N020 G00 X46. Z2.；
            N030 G01 Z-52.；
程序内容   N040 X48.625；
            N050 Z-60.；
            N060 X85.；
            N070 G00 X100.；
程序结束   N080 M30；
</pre>

程序号用作加工程序的开始标识。每个工件的加工程序都有自己专用的程序号。不同的数控系统，程序号地址码也不相同，常用的有 %、P、O 等符号，编程时一定要按照系统说明书的规定去指定，如写成 %8、P10、O0001 等形式，否则系统不识别。程序内容由加工顺序、刀具的各种运动轨迹和各种辅助动作的若干个程序段组成。结束符号表示加工程序结束，例如，FANUC 系统中用 M02 表示；若需程序返回至程序开始处，则需使用 M30 指令。

程序段中的各坐标数值输入时应至少带一位小数，每段程序最后应加 ";" 以示此段程序结束。

（2）程序段的格式。一个程序段定义一个将由数控装置执行的指令行。程序段的格式定义了每个程序段中功能字的句法，其结构如图 1-4-9 所示。

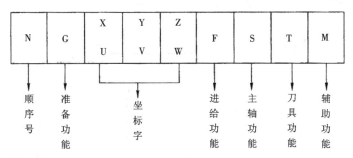

图 1-4-9　程序段的格式

（3）程序指令字符的格式。一个指令字符是由地址符（指令字符）和带符号（如定义尺寸的字）或不带符号（如准备功能字 G 代码）的数据组成的。程序中不同的指令字符及其后的数据确立了每个指令字符的含义，在数控程序段中包含的常用指令字符见表 1-4-3。

表 1-4-3　常用指令字符

功能	指令字符	意义
程序号	O	程序编号（0~9999）
程序段顺序号	N	程序段顺序号（N0~N…）
准备功能	G	指令动作方式（如直线、圆弧等）

续表

功能	指令字符	意义
尺寸字	X, Y, Z, D, V, W, A, B, C	移动坐标轴
	R	圆弧半径、固定循环的参数
	I, J, K	圆心坐标
进给功能	F	指定进给速度
主轴功能	S	指定主轴转速
刀具功能	T	选择刀具编号
辅助功能	M	机床开、关控制及其他相关控制
暂停	P, X	指定暂停时间
子程序号指定	P	指定子程序号
重复次数	L	子程序的重复次数
参数	P, Q, R, U, W, I, K, C, A	车削复合循环参数
倒角控制	C, R	自动倒角参数

四、任务实施

1. 典型零件的加工工艺

加工图 1-4-10 所示零件，长棒料，毛坯直径为 45 mm，要求一次装夹。

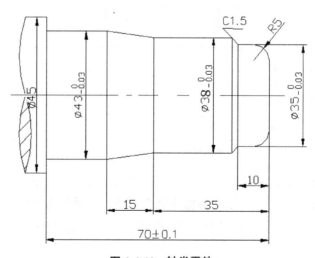

图 1-4-10　轴类零件

1）工艺分析

（1）零件外形复杂，需加工外圆、锥体、凸圆弧及倒角。

（2）根据零件形状选用如下刀具。

T01 外圆粗刀：加工余量大，要求副偏角不发生干涉。

T02 外圆精车刀：菱形刀片，刀尖圆弧 *R*0.4 mm，副偏角 >35°。

（3）坐标计算：根据选用的指令，此零件如用 G01、G02 指令编程，粗加工路线复杂，尤其圆弧处的计算和编程烦琐；适宜用 G71 指令，加工时依图形得出精车外形各坐标点，一次处理编程。

2）FANUC 数控系统工艺及编程路线

（1）1 号刀平端面。

（2）1 号刀用 G71 指令粗加工外形。

（3）2 号刀用 G70 指令精加工外形。

3）FANUC 数控系统参考程序

FANUC 数控系统加工轴类零件参考程序见表 1-4-4。

表 1-4-4　加工轴类零件程序

O1401;	程序名;
N1;	程序段号（粗加工段）;
G99 T0101 M03 S600;	换 1 号外圆刀，主轴正转，转速 600 r/min;
G00 X100. Z100. M08;	快速走到中间安全点，开冷却液;
G00 X47.Z2.;	循环起点;
G71 U1.5 R0.5 ;	外形复合循环加工，*X* 向背吃刀量 1.5 mm，退刀量 0.5 mm;
G71 P10 Q20 U0.5 W0.02 F0.2;	精加工程序段 N10~N20，*X* 向余量 0.5 mm，*Z* 向余量 0.02 mm;
N10 G00 X0.	精加工第一段;
G01 Z0 F0.1 ;	平端面;
G01 X25. ;	圆弧起点;
G03 X35.Z–5.R5;	加工凸圆;
G01 Z–10.;	加工 *ϕ*35 mm 外圆;
X38. C1.5;	倒角;
Z–35.;	加工 *ϕ*38 mm 外圆;
X43. W–15.;	加工锥体;
Z–70.;	加工 *ϕ*43 mm 外圆;
N20 G00 G40 X47.;	退刀;
G00 X100. Z100.;	回换刀点;
M05;	主轴停;
M09;	切削液停;
M00;	程序停;
N2;	精加工;
G99 M03 S800 T0202 ;	换 2 号刀精车外圆;
G00 X47. Z2. ;	循环起点;
G70 P10 Q20 F0.1;	精加工外形;
G00 X100. Z100.;	回换刀点;
M05;	主轴停;
M09;	切削液停;
M30;	程序停

五、任务评价

数控车床程序编制、校验考核。

（1）考核形式：实际操作。

（2）考核内容：

①编制程序；

②输入程序；

③校验程序。

（3）考核要求：能独立完成相关实际操作。

项目二

轴类工件编程与加工

【项目描述】

本项目内容是数控加工的基本内容之一，将涉及较多的数控编程及加工方面的知识。比如，需掌握轴类零件及刀具的装夹方法，会填写加工刀具卡和工艺卡，能够完成轴类零件编程与加工，熟悉轴类零件加工程序的检验方法。并且理论联系实际，在编程教室利用数控仿真软件进行仿真加工，进入实训车间现场进行机床加工，分小组协作完成或单独完成加工任务。

【学习目标】

（1）了解轴类零件的结构特点；

（2）对轴类零件进行数控车削工艺分析；

（3）熟练掌握用 G00、G01、G02、G03、G90、G71、G73、G70、G32、G92 指令编制轴类零件的加工程序；

（4）正确选择、使用加工轴类零件常用的刀具及切削用量；

（5）能够操作 FANUC 0i 系统数控车床完成零件的加工。

任务一　　台阶轴类零件编程与加工

一、图样与技术要求

轴类零件是机械行业中最重要、最常用的机械元件之一。本次任务着重讲解台阶轴类零件的图纸分析、加工工艺安排和编程加工。图 2-1-1 所示台阶轴是本次要完成的加工零件，毛坯尺寸为 ϕ44 mm×73 mm。

二、图纸分析与工艺安排

1. 图纸分析

图示 2-1-1 为台阶轴，形状简单，结构尺寸变化不大，有 3 个台阶面，径向尺寸中，ϕ40 mm、ϕ42 mm 两个尺寸公差为 0.05 mm，要求不高。轴向尺寸中 ϕ40 mm 外圆段有长度公差要求，表面结构参数值不大于 Ra3.2 μm。

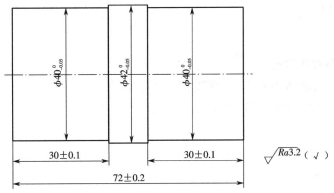

图 2-1-1 台阶轴

2. 工艺安排

1）确定工件的装夹方案

根据毛坯的形状，可选择三爪自动定心卡盘装夹，编程原点设在工件右端面与轴线交点处。此零件需经二次装夹，才能完成加工，第一次夹工件左端，车 $\phi42$ mm、$\phi40$ mm 外圆。第二次以 $\phi40$ mm 精车外圆为定位基准，平端面，确定总长，然后进行 $\phi40$ mm 外圆的加工。

2）确定加工路线

（1）平端面，定基准；

（2）粗、精车 $\phi42$ mm、$\phi40$ mm 外圆；

（3）工件调头，夹 $\phi40$ mm 外圆；

（4）平端面（确定总长）；

（5）粗、精车 $\phi40$ mm 外圆。

3）填写加工刀具卡和工艺卡

加工刀具卡和工艺卡，见表 2-1-1

表 2-1-1 加工刀具卡和工艺卡

零件图号	2-1-1	数控车床加工工艺卡		机床型号	CKA6150
零件名称	台阶轴			机床编号	
刀具表				量具表	
刀具号	刀具补偿号	刀具名称	刀具参数	量具名称	规格 /mm
T01	01	90°外圆车刀	C 型刀片	游标卡尺 千分尺	0~150/0.02 25~50/0.01
工序	工艺内容			切削用量			加工性质
				$S/$（r/min）	$F/$（mm/r）	a_p/mm	
数控车	车外圆，平端面，确定基准			500	—	1	手动
1	加工 $\phi42$ mm 外圆			600~1 000	0.1~0.2	0.5~3	自动
2	加工 $\phi40$ mm 外圆			600~1 000	0.1~0.2	0.5~3	自动
数控车	调头夹 $\phi40$ mm 外圆，平端面，保证工件总长			500	—	1	手动
1	加工 $\phi40$ mm 外圆			600~1 000	0.1~0.2	0.5~3	自动

三、程序编制及加工

1. G00 快速点定位指令

G00 指令能快速移动刀具到达指定的坐标点位置，用于刀具进行加工以前的空行程移动或加工完成的快速退刀；G00 指令使刀具快速运动到指定点，以提高加工效率，不能进行切削加工。

指令格式：G00 X（U）_ Z（W）_;

指令说明如下。

（1）绝对值编程：G00 X_Z_ 表示终点位置相对于工件原点的坐标值，轴向移动方向由 Z 轴坐标值确定，径向进退刀时在不过轴线情况下都为正值。如：两轴同时移动 G00 X80. Z10.，单轴移动 G00 X50. 或 G00 Z-10. 。

（2）增量值编程：G00 U_ W_；U_ W _ 表示刀具从当前所在点到终点的距离和方向；U 表示直径方向移动量，即大、小直径量之差，W 表示移动长度，U、W 移动方向都由正、负号确定。计算 U、W 移动距离的起点坐标值是执行前一程序段移动指令的终点值。也可在同一移动指令里采用混合编程。如：G00 U20. W30.，G00 U-5. Z40. 或 G00 X80. W40.。

（3）用 G00 指令编程时，也可以省略一个"0"，写作 G0。

例：如图 2-1-2 所示刀具要快速移到指定位置，用 G00 编写程序段。

绝对值方式编程：G00　X50.0　Z6.0;

增量值方式编程：G00　U-70.0　W-84.0;

2. G01 进给切削指令

G01 又称直线插补功能，该指令用于使刀具以指定的进给速度移动到指定的位置。当主轴转动时，可用于对工件以一定的速度进行切削加工。

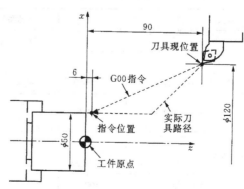

图 2-1-2　G00 走刀路径

指令格式：G01 X（U）_ Z（W）_;

指令说明如下。

（1）绝对值编程：G01 X_Z_；表示终点位置相对于工件原点的坐标值，轴向移动方向由 Z 轴坐标值确定，径向进退刀时在不过轴线情况下都为正值。如：两轴同时移动 G01 X80. Z10.，单轴移动 G01 X50. 或 G01 Z-10. 。

（2）增量值编程：G01 U_ W_；U_W_ 表示刀具从刀具当前所在点到终点的距离和方向；U 表示直径方向移动量，即大、小直径量之差，W 表示移动长度，U、W 移动方向都由正、负号确定。计算 U、W 移动距离的起点坐标值是执行前一程序段移动指令的终点值，也可在同一移动指令里采用混合编程。如：G01 U20. W30.，G01 U-5. Z40. 或 G01 X80. W40.。

（3）机床在执行 G01 指令时，在该程序段中必须具有或在该程序段前已经有 F 指令，否则系统认为进给速度为零。

（4）用 G01 指令车削如图 2-1-3 所示零件时刀具运动轨迹如图中所示。

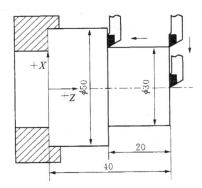

图 2-1-3 用 G01 指令车外圆刀具运动轨迹

3. 编程举例

用 G01 指令加工如图 2-1-4 所示的零件，加工程序见表 2-1-2。

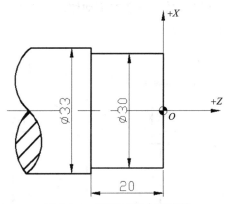

图 2-1-4 用 G01 指令车外圆

表 2-1-2　用 G01 指令车外圆程序

程序内容	程序说明
O2001；	程序号；
N010 G99 M03 S600 T0101；	进给速度 mm/r，主轴正转速度 600 r/min，选 1 号刀；
N020 G00 X100.Z100.；	换刀点；
N030 G00 X30.5 Z2.；	对刀点；
N040 G01 X30.5 Z−20. F0.2；	粗车 ϕ30 mm 外圆，留 0.5 mm 精加工余量；
N050 G01 X34. Z−20.；	X 轴方向退刀；
N060 G00 X34. Z2. S1000；	Z 轴方向退刀，给出精加工转速 1 000 r/min；
N070 G01 X30. Z2. F0.1；	精车 ϕ30 mm 外圆起点；
N080 G01 X30. Z−20.；	精车 ϕ30 mm 外圆终点；
N090 G01 X34. Z−20.；	X 轴方向退刀；
N100 G00 X100. Z100.；	返回换刀点；
N110 M05；	主轴停；
N120 M30；	程序结束返回程序头

4. 轴类零件加工编程的单一循环指令（G90）

1）指令格式

G90 X（U）_Z（W）_F_；

说明：

X、Z——切削圆柱面的终点坐标值；

U、W——切削圆柱面的终点相对于循环起点坐标分量；

F——进给速度。

2）单一固定循环切削加工路线

图 2-1-5 所示为单一固定循环切削加工路线，刀具从循环起点开始按照矩形循环，最后回到循环起点，图中虚线表示按快进速度 R 运动，实线表示按工作进给速度 F 移动。

单一固定循环指令可以将一系列连续加工动作，如"切入—切削—退刀—返回"，用一个循环指令完成，从而简化程序。

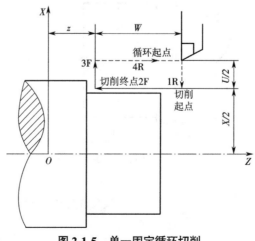

图 2-1-5　单一固定循环切削

注意：使用循环切削指令时，刀具必须先定位至循环起点，再执行循环切削指令，且完成一轮循环切削后，刀具仍回到此循环起点。

3）编程举例

用 G90 指令加工如图 2-1-6 所示零件，加工程序见表 2-2-3。

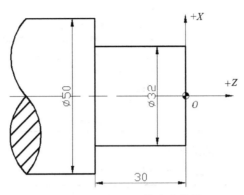

图 2-1-6 用 G90 指令车外圆

表 2-1-3 用 G90 指令车外圆程序

程序内容	程序说明
O2201; N010 G99 M03 S600 T0101; N020 G00 X100. Z100.; N030 G00 X55.0 Z5.0;	程序号; 进给速度 mm/r，主轴正转速度 600 r/min，选 1 号刀; 换刀点; 循环起点;
N040 G90 X46.0 Z-30. F0.2; N050 X42.; N060 X38.; N070 X34.; N080 X32.4; N090 X32.0 Z-30.0 S1000 F0.1; M100 G00 X100. Z100.; N110 M05; N120 M30;	粗车 $\phi32$ mm 外圆，第一刀，切深为 4 mm; 粗车 $\phi32$ mm 外圆，第二刀，切深为 4 mm; 粗车 $\phi32$ mm 外圆，第二刀，切深为 4 mm; 粗车 $\phi32$ mm 外圆，第四刀，切深为 4 mm; 粗车 $\phi32$ mm 外圆，第五刀，留 0.4 mm 的精车余量; 精车 $\phi32$ mm 外圆; 返回换刀点; 主轴停; 程序结束，返回程序头

5. 零件加工程序

根据图 2-1-1 所示零件，分析工件加工路线，确定加工装夹方案以及采用的刀具和切削用量，根据工艺过程将工序内容划分为两个部分，并对应编制两个程序完成加工任务。

表 2-1-4 为加工台阶轴右端的程序。

表 2-1-4　加工台阶轴右端程序

程序内容	程序说明
O2202;	程序号;
N010 G99 M03 S600 T0101;	进给速度 mm/r, 主轴正转 600 r/min, 选 1 号刀;
N020 G00 X100.Z100.;	换刀点;
N030 G00 X45. Z2.;	循环起点;
N040 G90 X42.5 Z-42. F0.2;	粗车 $\phi42$ mm 外圆, 留 0.5 mm 精加工余量;
N050 X40.5 Z-30.;	粗车 $\phi42$ mm 外圆, 留 0.5 mm 精加工余量;
N060 X40. Z-30. S1000 F0.1;	精车 $\phi42$ mm 外圆, 给出精车转数和进给量;
N070 X42. Z-42.;	精车 $\phi42$ mm 外圆;
N080 G00 X100. Z100.;	返回换刀点;
N090 M05;	主轴停;
N100 M30;	程序结束, 返回程序头

表 2-1-5 为加工台阶轴左端的程序。

表 2-1-5　加工台阶轴左端程序

程序内容	程序说明
O2203;	程序号;
N010 G99 M03 S600 T0101;	进给速度 mm/r, 主轴正转速度 600 r/min, 选 1 号刀;
N020 G00 X100.Z100.;	换刀点;
N030 G00 X45. Z2.;	循环起点;
N040 G90 X40.5 Z-30. F0.2;	粗车 $\phi40$ mm 外圆, 留 0.5 mm 精加工余量;
N050 X40. Z-30. S1000 F0.1;	精车 $\phi40$ mm 外圆, 给出精车转数和进给量;
N080 G00 X100. Z100.;	返回换刀点;
N090 M05;	主轴停;
N100 M30;	程序结束, 返回程序头

6.零件加工

（1）教师演示工件加工准备工作及加工过程。

（2）教师讲解演示尺寸公差的保证方法。

（3）学生分组加工工件，教师巡回指导，及时发现并纠正学生加工中出现的问题。

四、任务评价

按表 2-1-6 所示各项内容进行任务评价。

表 2-1-6　加工台阶轴综合评分标准

	考核项目	考核要求	配分	评分标准	检测结果	得分	备注
1	外圆尺寸	$\phi42_{-0.05}^{0}$ mm	10	每超差 0.01 mm 扣 5 分			
2		$\phi40_{-0.05}^{0}$ mm	15×2	每超差 0.01 mm 扣 5 分			
3	端面结构参数值	Ra3.2 μm（两处）	5×2	超差全扣			
4	倒钝角	四处	5×4	超差全扣			

5	长度	30 mm、30 mm、72 mm	5×3	超差全扣			
6	表面结构参数值	Ra1.6 μm（三处）	5×3	超差全扣			
7	工艺、程序	遵守工艺与程序的有关规定		违反规定扣总分 1~5 分			
8	规范操作	遵守数控车床规范操作的有关规定		违反规定扣总分 1~5 分			
9	安全文明生产	遵守安全文明生产的有关规定		违反规定扣总分 1~5 分			

学生在任务实施过程的小结及反馈：

教师点评：

任务二　圆弧与锥体零件编程与加工

一、图样与技术要求

本任务需加工的零件由凹凸圆弧、圆锥面、圆柱面等组成，相对较为复杂，指令选择、程序编写、车削路线的设计较困难，主要原因是在半精车凹圆弧时一次性去除的余量较大，易引起打车刀等事故。为了解决这些问题。本项任务将介绍 G73 指令——封闭切削粗车循环指令的使用。

二、图纸分析与工艺安排

1. 图纸分析

图 2-2-1 所示为圆弧与锥体零件，毛坯材料尺寸为 ϕ50 mm×220 mm，要求按图样单件加工。

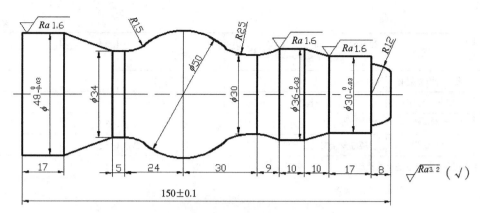

图 2-2-1　圆弧与锥体零件

2. 工艺安排

1）确定工件装夹方案

根据毛坯形状，可选择三爪自动定心卡盘装夹，编程原点设在工件右端面与轴线交点处。此零件只需经一次装夹，就能完成加工。

2）确定加工路线

（1）平端面；

（2）按照工件轮廓使用 G73 循环指令任务可完成加工任务。

3）填写加工刀具卡和工艺卡

加工刀具卡和工艺卡，见表 2-2-1。

表 2-2-1　加工刀具卡和工艺卡

零件图号	2-2-1	数控车床加工工艺卡		机床型号	CKA6150
零件名称	圆弧与锥体零件			机床编号	
刀具表				量具表	
刀具号	刀具补偿号	刀具名称	刀具参数	量具名称	规格 /mm
T01	01	93°外圆车刀	D 型刀片	游标卡尺 千分尺	0～150/0.02 25～50/0.01
工序	工艺内容			切削用量			加工性质
				S/（r/min）	F/（mm/r）	a_p/mm	
数控车	车外圆、端面，确定基准			500	—	1	手动
1	车外圆、端面，确定基准			500	—	1	手动
2	加工 $R12$ mm 圆弧			600～1 000	0.1～0.2	0.5～3	自动
3	加工 $\phi30$ mm 外圆			600～1 000	0.1～0.2	0.5～3	自动
4	加工圆锥			600～1 000	0.1～0.2	0.5～3	自动
5	加工 $\phi36$ mm 外圆			600～1 000	0.1～0.2	0.5～3	自动
6	加工 $R25$ mm 圆弧			600～1 000	0.1～0.2	0.5～3	自动
7	加工 $\phi50$ mm 圆弧			600～1 000	0.1～0.2	0.5～3	自动
8	加工 $\phi34$ mm 外圆			600～1 000	0.1～0.2	0.5～3	自动
9	加工圆锥			600～1 000	0.1～0.2	0.5～3	自动

三、程序编制及加工

1. 圆锥计算

圆锥是车削加工中常见的零件形式之一，如图 2-2-2 所示，常用的参数有，圆锥大端直径 D、圆锥小端直径 d、圆锥长度 L、锥度比 C，它们之间的关系为：

$$C=(D-d)/L$$

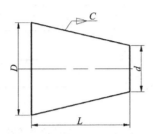

图 2-2-2   圆锥计算常用参数

2. 粗车复合循环指令（G71 指令）

在使用 G71 指令时只需在程序中指定精加工路线，给出粗加工每次吃刀量，指令就会自动重复切削，配合 G70 指令精加工循环，直至完成零件的加工，相对于 G01 和 G00，G71 指令使编程变得简便，程序内容也大为缩短，适用于车削圆棒料毛坯的零件。

1）指令格式

G71 U（Δd）R（e）；

G71 P（ns）Q（nf）U（Δu）W（Δw）F（f）S（s）T（t）；

说明：

$\triangle d$——X 轴向每次切削深度（半径值）；

e——退刀量；

ns——精加工形状程序的第一个程序段的段号；

nf——精加工形状程序的最后一个程序段的段号（终点为 B 点的程序段）；

$\triangle u$—— X 轴方向上的精加工余量（直径值）；

$\triangle w$—— Z 轴方向上的精加工余量；

f、s、t——包含在 ns 到 nf 程序段中的任何 F、S 或 T 功能在循环中被忽略，而在 G71 程序段中的 F、S 或 T 功能有效。

2）G71 指令车削路线及参数示意图

G71 指令段中的参数如图 2-2-3 所示。首先根据用户编写的精车加工路线和每次切削深度，在预留出 X 轴方向和 Z 轴方向精加工余量后，计算出粗加工的刀数和每刀的路线坐标值，刀具按层以加工外圆柱面的形式将余量切除，然后形成与精加工轮廓相似的轮廓。粗加工结束后，可使用 G70 指令完成精加工。

如图 2-2-3 所示，刀具起始点为 A，此指令可实现背吃刀量为 Δd，精加工余量为 $\Delta u / 2$ 和 Δw 的粗加工循环。其中 Δd 为背吃刀量（半径值），该量无正负号，刀具的切削方向取决于 AA' 方向；e 为退刀量，可由参数设定；ns 指定精加工路线的第一个程序段的顺序号；nf 指定精加工路线的最后一个程序段的顺序号。

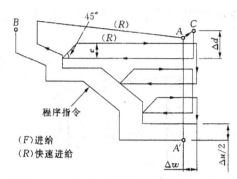

图 2-2-3　G71 指令车削路线及参数示意图

3. 精车循环指令（G70 指令）

1）指令格式

G70　P（ns）　Q（nf）；

说明：

ns——精加工程序第一个程序段的段号；

nf——精加工程序最后一个程序段的段号。

例如：G70　P10　Q20；

2）G70 指令的车削轨迹

G70 指令车削工件时，刀具沿着工件的实际轮廓进行切削，循环结束后刀具自动返回循环起点。

注意：G70 指令是精车循环指令，因此在使用时需要配合粗车循环指令（G71、G72、G73），不能单独使用。

G70 指令执行过程中的 F、S 值，由段号 "ns" 和 "nf" 之间的 F、S 指定。

3）编程举例

零件如图 2-2-4 所示，毛坯材料尺寸为 $\phi45$ mm×78 mm。

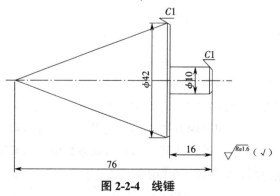

图 2-2-4　线锤

根据图 2-2-4 所示零件，分析工件加工路线，确定加工装夹方案以及加工采用的刀具和切削用量，根据工艺过程将工序内容划分为两个部分，并对应编制两个程序，完成加工任务。表 2-2-2 为加工线锤右端的程序，表 2-2-3 为加工线锤左端的程序。

表 2-2-2 加工线锤右端的程序

程序内容	程序说明
O2902；	程序名；
N010 G99 M03 S600 T0101；	进给速度 mm/r，主轴正转速度 600 r/min，选 1 号刀；
N020 G00 X100. Z100.；	换刀点；
N030 G00 X50. Z2.；	循环起点；
N040 G71 U2. R0.5；	给出切削深度 2 mm，退刀量 0.5 mm；
N050 G71 P60 Q70 U0.5 W0 F0.2；	精车路线为 N060—N070；
N060 G00 X0；	
G01 Z0 F0.1；	
X8.；	
X10. Z-1.；	
Z-16.；	
X41.37；	
X42.37 Z-16.5；	
N070 X45.；	退刀；
N090 G70 P60 Q70；	精车；
N100 G00 X100. Z100.；	返回换刀点；
N110 M05；	主轴停；
N120 M30；	程序结束，返回程序头

表 2-2-3 加工线锤左端的程序

程序内容	程序说明
O2903；	程序名；
N010 G99 M03 S600 T0101；	进给速度 mm/r，主轴正转速度 600 r/min，选 1 号刀；
N020 G00 X100. Z100.；	换刀点；
N030 G00 X50. Z2.；	循环起点；
N040 G71 U2. R0.5；	给出切削深度 2 mm，退刀量 0.5 mm；
N050 G71 P60 Q70 U0.5 W0 F0.2；	精车路线为 N060—N070
N060 G00 X0；	
G01 Z0 F0.1；	
X42.37 Z60.5；	
N070 X45.；	退刀；
N080 G70 P60 Q70；	精车；
N090 G00 X100. Z100.；	返回换刀点；
N100 M05；	主轴停；
N110 M30；	程序结束，返回程序头

4. 封闭切削粗车循环指令（G73 指令）

G73 指令主要用于按固定轨迹切削工件轮廓，这种复合循环可以高效切削锻造成形、铸造成形或已经粗车成形的工件。对不具备类似成形条件的工件，采用该指令反而会增加刀具在切削过程中的空行程，降低加工效率。

1）指令格式

G73 U（Δi）W（Δk）R（d）；

G73 P（ns）Q（nf）U（Δu）W（Δw）F（f）S（s）T（t）；

说明：

Δi——粗切时径向切除的总余量；

Δk——粗切时轴向切除的总余量 ；

d ——循环次数。

其他参数含义与 G71 指令相同。

2）G73 指令的车削路线

G73 指令走刀路线如图 2-2-5 所示。执行 G73 指令时，每一刀切削路线的轨迹形状是相同的，只是位置不同。每走完一刀，就把切削轨迹向工件移动一个位置，因此对于经锻造、铸造等粗加工已初步成形的毛坯，可高效加工。

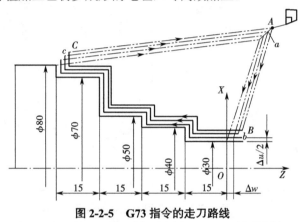

图 2-2-5　G73 指令的走刀路线

3）编程举例

加工图 2-2-6 所示零件。加工过程中需注意以下两个问题。

（1）刀具选择。加工该零件时首先要注意选用刀具的副偏角要大，目的在于加工过程中，避免过切，例如可选用尖刀。

（2）加工程序见表 2-2-4。基点计算，为了简便，可采用 CAD 制图捕捉基点的方法，如图 2-2-7 所示。

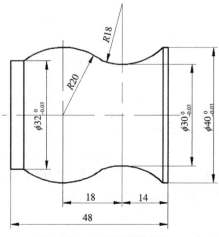

图 2-2-6　凹凸圆弧零件加工

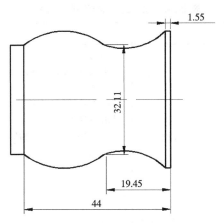

图 2-2-7　CAD 捕捉基点

表 2-2-4　凹凸圆弧零件加工程序

程序内容	程序说明
O2111;	程序号;
N010 G99 M03 S600 T0101 ;	进给速度 mm/r，主轴正转速度 600 r/min，选 1 号刀;
N020 G00 X100. Z100.;	换刀点;
N030 G00 X47. Z5.	循环起点;
N040 G73 U7.5 R5;	切深 2 mm，退刀 0.5 mm ;
N050 G73 P60 Q70 U1.0 W0.1 F0.2 ;	精车 N60 至 N070，精车余量 X 向 1mm，Z 向 0.1 mm，接近工件;
N060 G00 X0 ;	平端面起点 ;
G01 Z0　F0.1 ;	平端面;
X40.0 ;	
Z−1.55 ;	加工 $\phi40$ mm 外圆;
G02 X32.11 Z−19.45 R18. ;	加工 R18 mm 圆弧;
G03 X32. Z−44. R20. ;	加工 R20 mm 圆弧;
G01 Z−48.;	加工 $\phi32$ mm 外圆;
N070　X47. ;	退刀;
N080 G70 P60 Q70 ;	精加工指令;
N090　X100.0 Z100.0;	返回换刀点;
N100 M05 ;	主轴停;
N110 M30;	主程序结束并返回

5. 零件加工程序

分析图 2-2-1 所示零件，确定工件加工路线及加工装夹方案，选用刀具和确定切削用量，根据工艺过程将工序内容划分为一个部分。

表 2-2-5 为加工图 2-2-1 所示圆弧与锥体零件的程序。

表 2-2-5　加工圆弧与锥体零件程序

程序内容	程序说明
O2112；	程序号；
N010 G99 M03 S600 T0101；	进给速度 mm/r，主轴正转速度 600 r/min，选 1 号刀；
N020 G00 X100. Z100.；	换刀点；
N030 G00 X62.0 Z5.0	循环起点；
N040 G73 U21. R11；	切削深度 4 mm；
N050 G73 P060 Q070 U1.0 W0.1 F0.2 ；	精车 N060 至 N070，精车余量 X 向 1 mm，Z 向 0.1 mm；
N060 G00 X17.89 Z2.；	加工轮廓起点；
G01　　　Z0 F0.1 ；	
G03 X24.　Z−8. R12；	
G01 X30.；	
Z−25.；	
X36.　Z−35.；	
Z−45.；	
G02 X32.83 Z−55.41 R25.；	
G03 X39.12 Z−89.84 R25.；	
G02 X34.　Z−108 R15.；	
G01　　　Z−113；	
X48.　Z−133.；	
Z−150.；	
N070 G00 X62.0 ；	
N080 G70 P060 Q070 ；	精加工指令；
N090 X100.0 Z100.0；	返回换刀点；
N100 M05 ；	主轴停；
N110 M30；	主程序结束并返回

6. 零件加工

（1）教师演示加工工件的准备工作及加工过程。

（2）教师讲解、演示保证尺寸公差的方法。

（3）学生分组完成工件的加工任务，教师巡回指导，及时发现并纠正学生加工操作中出现的问题。

四、任务评价

按表 2-2-6 所示各项内容进行任务评价。

表 2-2-6　圆弧与锥体零件评分标准

	考核项目	考核要求	配分	评分标准	检测结果	得分	备注
1	外圆	$\phi30_{-0.03}^{0}$ mm	7	每超差 0.01 mm 扣 5 分			
2		$\phi36_{-0.03}^{0}$ mm	7	每超差 0.01 mm 扣 5 分			
3		$\phi48_{-0.03}^{0}$ mm	7	每超差 0.01 mm 扣 5 分			
4		$\phi34$ mm	7	每超差 0.01 mm 扣 5 分			

续表

	考核项目	考核要求	配分	评分标准	检测结果	得分	备注
5	长度	8 mm、17 mm、10 mm（两处）、9 mm、30 mm、24 mm、5 mm、17 mm、150 mm	2×10	超差全扣			
6	圆弧	R12 mm	5	超差全扣			
7		R25 mm	5	超差全扣			
8		R15 mm	5	超差全扣			
9		ϕ50 mm	5	降一级扣2分			
10	圆锥	两处	4×2	超差全扣			
11	端面	两处	2×2	超差全扣			
12	表面结构参数值	Ra1.6 μm（三处）	2×3	降一级扣2分			
		Ra1.6 μm（七处）	2×7	降一级扣2分			
13	工艺、程序	遵守工艺与程序的有关规定		违反规定扣总分1~5分			
14	规范操作	遵守数控车床规范操作的有关规定		违反规定扣总分1~5分			
15	安全文明生产	遵守安全文明生产的有关规定		违反规定扣总分1~5分			

学生任务实施过程的小结及反馈：

教师点评：

任务三　　三角螺纹零件编程与加工

一、图样与技术要求

　　螺纹是零件上常见的一种结构，带螺纹的零件是机器设备中重要的零件之一。作为标准件，它的用途十分广泛，能起到连接、传动、紧固等作用。螺纹按用途可分为连接螺纹和传动螺纹两种。在 FANUC 系统数控车床上加工螺纹可用 G32 指令、G92 指令来进行编程，但每种编程方法都有自己的特点。本任务以图 2-3-1 所示螺柱零件为例，重点讲解螺纹部分的编程方法。

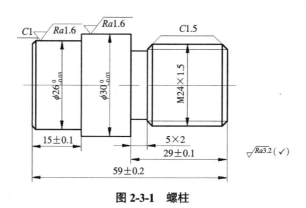

图 2-3-1　螺柱

二、图纸分析

1. 图纸分析与工艺安排

　　图 2-3-1 所示零件为螺柱，是一种比较典型的螺纹件。本次加工任务为主要完成外圆、沟槽、三角形圆柱外螺纹的加工，加工过程中重点掌握螺纹加工的相关知识及有关编程指令。

2. 工艺安排

1）确定工件的装夹方案和加工路线

根据毛坯的形状，可选择三爪自动定心卡盘装夹，编程原点设在工件右端面与轴线交点处。

此零件需经二次装夹才能完成加工。第一次夹工件右端面，车 $\phi30$ mm 外圆、$\phi26$ mm 外圆；第二次装夹以 $\phi26$ mm 精车外圆为定位基准，平端面，确定总长，车 $\phi24$ mm 外圆和 5 mm×2 mm 的退刀槽，然后进行 M24×1.5 的三角形圆柱外螺纹加工。

2）填写加工刀具卡和工艺卡

加工刀具卡和工艺卡，见表 2-3-1。

表 2-3-1 加工刀具卡和工艺卡

零件图号	2-3-1	数控车床加工工艺卡		机床型号	CKA6150
零件名称	螺柱			机床编号	
刀具表				量具表	
刀具号	刀具补偿号	刀具名称	刀具参数	量具名称	规格 /mm
T01	01	93° 外圆车刀	D 型刀片	游标卡尺 千分尺	0~150/0.02 25~50/0.01
T02	02	切槽刀	刀宽 4 mm	游标卡尺 千分尺	0~150/0.02 25~50/0.01
T03	03	螺纹刀	60° 三角形外螺纹刀	游标卡尺	0~150/0.02

工序	工艺内容	切削用量			加工性质
		S/（r/min）	F/（mm/r）	a_p/mm	
数控车	车外圆、端面确定基准	500	—	1	手动
1	加工 $\phi30$ mm 外圆	600~1 000	0.1~0.2	0.5~3	自动
2	加工 $\phi26$ mm 外圆	600~1 000	0.1~0.2	0.5~3	自动
数控车	调头夹 $\phi26$ mm 外圆，平端面，保总长	500	—	1	手动
1	加工 $\phi24$ mm 外圆	600~1 000	0.1~0.2	0.5~3	自动
2	加工 5 mm×2 mm 的退刀槽	600~1 000	0.1~0.2	0.5~3	自动
3	加工 M24×1.5 的三角形圆柱外螺纹	600~800	1.5	—	自动

三、程序编制及加工

1. 槽类工件的装夹方案

根据槽的宽度条件，在切槽时经常采用直接成型法，就是说槽的宽度就是切槽刀刀刃的宽度，也就等于背吃刀量 a_p。用这种方法切削时会产生较大的切削力。另外，大多数槽是位于零件的外表面上的，切槽时主切削刃的方向与工件轴线平行，会影响到工件的装夹稳定性。因此，在数控车床上进行槽加工一般可采用下面两种装夹方式。

（1）利用软卡爪装夹，并适当增加夹持面的长度，以保证定位准确、装夹稳固。

（2）利用尾座及顶尖做辅助支撑，采用一夹一顶方式装夹，最大限度保证零件装夹稳定。

2. 刀具的选择与切槽的方法

1）切槽刀的选择

切槽时常选用高速钢切槽刀和机夹可转位切槽刀。切槽刀的几何形状和角度如图2-3-2所示。选择切槽刀时要注意两点：一是切槽刀的宽度 a 要适宜，二是切削刃长度 L 要大于槽深。图2-3-3所示为机夹可转位外切槽刀。

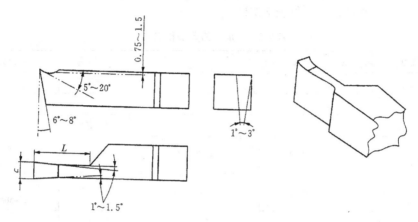

图 2-3-2　高速钢切槽刀

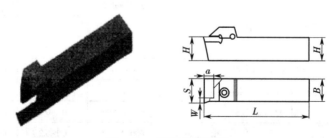

图 2-3-3　可转位切槽刀

2）切槽的方法

（1）对于宽度、深度值不大，且精度要求不高的槽，可采用与槽等宽的刀具直接切入一次成型，如图2-3-4所示。刀具切入到槽底后可利用延时指令使刀具短暂停留，以修整槽底圆度，退出过程中可采用工进速度。

（2）对于宽度值不大，但深度值较大的深槽零件，为了避免切槽过程中由于排屑不畅，使刀具前部压力过大出现扎车刀和折断刀具的现象，应采用分次进刀的方式，即刀具在切入工件一定深度后，停止进刀并回退一段距离，达到断屑和排屑的

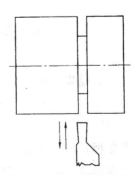

图 2-3-4　简单槽类零件加工方式

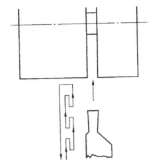

图 2-3-5 深槽零件加工方式

目的，如图 2-3-5 所示。同时注意尽量选择强度较高的刀具。

（3）宽槽的切削。通常把大于一个切刀宽度的槽称为宽槽，宽槽的宽度、深度的精度要求及表面质量要求相对较高。在切削宽槽时常采用排刀的方式进行粗切，然后用精切槽刀沿槽的一侧切至槽底，精加工槽底至槽的另一侧，再沿侧面退出，切削方式如图 2-3-6 所示。

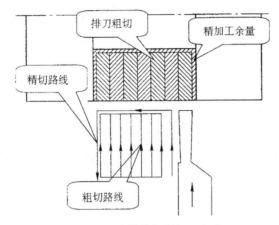

图 2-3-6 宽槽的切削加工方式

3. 切削用量与切削液的选择

背吃刀量、进给量和切削速度是切削用量三要素，在切槽过程中，背吃刀量受到切刀宽度的影响，其大小的调节范围较小。要增加切削稳定性，提高切削效率，就要选择合适的切削速度和进给速度。在普通车床上进行切槽加工，切削速度和进给速度相对外圆切削要选取得较低，一般取外圆切削速度的 30%~70%。数控车床的各项性能指标要远高于普通车床，在切削用量的选取上同样可以选择相对较高的速度，切削速度可以选择外圆切削速度的 60%~80%，进给速度选取 0.05~0.3 mm/r。

需要注意的是在切槽中容易产生振动现象，这往往是由于进给速度过低，或者是由于线速度与进给速度搭配不当造成的，需及时调整，以保证切削稳定。

切槽过程中，为了解决切槽刀刀头面积小、散热条件差、易产生高温而降低刀片切削性能等问题，可以选择冷却性能较好的乳化类切削液进行喷注，使刀具充分冷却。

4. 切槽（切断）编程指令

对于一般的单一切直槽或切断，采用 G01 指令即可，对于宽槽或多槽加工可采用子程序及复合循环指令进行编程加工。

G01 指令切槽编程举例：如图 2-3-7 所示，切削直槽，槽宽 5 mm 并完成两个 0.5 mm 宽的倒角，切槽刀宽为 4 mm。

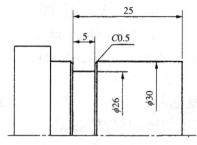

图 2-3-7 G01 指令切槽

（1）编程路线及过程：工件原点设在右端面，切槽刀对刀点为左刀位，因切槽刀宽小于槽宽，且需用切槽刀切倒角，故加工此槽需三刀完成。加工路线如图 2-3-8 所示。

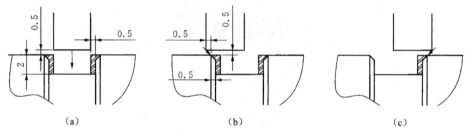

（a）   （b）   （c）

图 2-3-8 G01 指令切槽步骤示意

（2）各步骤程序如下。

如图 2-3-8（a）所示，先从槽中间将槽切至槽底并反向退出，左刀点位 Z 向坐标应为 -24.5 mm。

N010 T0202 M03 S500;

N020 G00 X31. Z–24.5;

N030 G01 X26. F0.05;

N040 X31.;

如图 2-3-8（b）所示，倒左角并切槽左边余量后退出。刀具起点设在倒角延长线上，应 X 向增加 0.5 mm 空距，Z 向也是 0.5 mm 空距，左刀点应往左移动边余量 0.5 mm+ 倒角宽 0.5 mm+ 起点延长 0.5 mm=1.5 mm。

N050 W–1.5;

N060 X29. W1;

N070 X26.;

N080 W0.5;

N090 X31.;

如图 2-3-8（c）所示，倒右角并切槽右边余量后移至槽中心退出，刀具应往右移动 1.5 mm。

N100 W1.5;

N110 X29. W−1.;

N120 X26.;

N130 W−0.5;

N140 X31.;

N150 G00 X100. Z100.;

N160 M05 M30;

5. 螺纹工件的相关知识

利用数控车床加工螺纹时，由数控系统控制螺距的大小和精度，从而简化计算，不用手动更换挂轮，并且螺距精度高不会出现乱扣现象；螺纹切削回程期间车刀快速移动，切削效率大幅提高；专用数控螺纹切削刀具、较高的切削速度的选用，又进一步提高了螺纹的形状精度和表面质量。

在螺纹加工中，背吃刀量 a_p 等于螺纹车刀切入工件表面的深度，随着螺纹刀的每次切入，背吃刀量在逐步增加。受螺纹牙型截面大小和深度的影响，螺纹切削的背吃刀量可能是非常大的，所以必须合理选择，见表 2-3-2。

表 2-3-2　常用螺纹切削的背吃刀量与切削次数　　　　　　　单位：mm

米制螺纹							
螺距 p/mm	1.0	1.5	2.0	2.5	3.0	3.5	4.0
牙深 h/mm	0.649	0.974	1.299	1.624	1.949	2.273	2.598
背吃刀量与切削次数 1次	0.7	0.8	0.9	1.0	1.2	1.5	1.5
2次	0.4	0.6	0.6	0.7	0.7	0.7	0.8
3次	0.2	0.4	0.6	0.6	0.6	0.6	0.6
4次		0.16	0.4	0.4	0.4	0.6	0.6
5次			0.1	0.4	0.4	0.4	0.4
6次				0.15	0.4	0.4	0.4
7次					0.2	0.2	0.4
8次						0.15	0.3
9次							0.2

注：表中给出的背吃刀量与切削次数为推荐值，编程者可根据自己的经验和实际情况进行选择。

6. 螺纹切削循环编程指令（G92 指令）

G92 指令可循环加工圆柱螺纹和圆锥螺纹，应用方式与 G90 外圆循环指令有类似之处。

1）指令格式

G00 X_ Z_（循环起点）

G92 X（U）_ Z（W）_ F_

说明：

X_ Z_——螺纹切削终点坐标值；

U、W——螺纹切削终点相对于循环起点的坐标增量；

F——螺纹的导程，单线螺纹时为螺距。

2）切削路线

如图 2-3-9 所示，切削路线如下：从循环起点快速移至螺纹起点（由循环起点坐标值和切削终点坐标值决定）—螺纹切削至螺纹终点—X 向快速退刀—Z 向快速回循环起点。

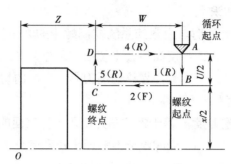

图 2-3-9　圆柱螺纹切削循环走刀路线

3）编程举例

按图 2-3-9 所示切削路线，程序见表 2-3-3。

表 2-3-3　圆柱螺纹切削循环程序

程序内容	程序说明
N3;	第三程序段号（螺纹加工段）；
N010 G95 M03 S600 T0303	进给速度设置 mm/r，主轴正转速度，600 r/min，选 3 号刀；
N020 G00 X32 Z3	循环起点；
N030 G92 X29.1 Z−22 F2	螺纹切削循环 1，进 0.9 mm；
N040　　X28.5 Z−22	螺纹切削循环 2，进 0.6 mm；
N050　　X27.9 Z−22	螺纹切削循环 3，进 0.6 mm；
N060　　X27.5 Z−22	螺纹切削循环 4，进 0.4 mm；
N070　　X27.4 Z−22	螺纹切削循环 5，进 0.1 mm；
N080 X100 Z100	返回换刀点；
N090 M05	主轴停；
N100 M30	程序结束，返回程序头

7. 零件加工程序

根据图 2-3-1 所示零件，确定加工装夹方案、工件加工路线以及采用的刀具和切削用量，根据工艺过程将工序内容划分为两个部分，并对应编制两个程序来完成加工任务。

表 2-3-4 为加工螺柱零件左端的程序。

表 2-3-4　加工螺柱零件左端的程序

程序内容	程序说明
O1111;	程序号;
N010 G99 M03 S600 T0101;	进给速度 mm/r, 主轴正转速度 600 r/min, 选 1 号刀;
N020 G00 X100.Z100.;	换刀点;
N030 G00 X30.5 Z2.;	接近工件点;
N040 G01 X30.5 Z–30. F0.2;	粗车 φ30 mm 外圆, 留 0.5 mm 精加工余量;
N050 G01 X32. Z–30.;	X 向退刀;
N060 G00 X32.Z2.;	Z 向退刀;
N070 G01 X28.5 Z2. F0.2;	粗车 φ26 mm 外圆第一刀起点;
N080 G01 X28.5 Z–30.;	粗车 φ26 mm 外圆第一刀终点;
N090 G01 X32. Z–30.;	X 向退刀;
N100 G00 X32. Z2.;	Z 向退刀;
N110 G00 X26.5 Z2.;	粗车 φ26 mm 外圆第二刀起点;
N120 G01 X26.5 Z–30. F0.2;	粗车 φ26 mm 外圆第二刀终点;
N130 G01 X32. Z–30.;	X 向退刀;
N140 G00 X32. Z2. S1000;	X 向退刀, 精加工转速 1 000 r/min;
N150 G01 X24. Z2.;	接近工件点;
N160 G01 X24. Z0. F0.1;	倒角起点;
N170 G01 X26. Z–1.;	倒角终点;
N180 G01 X26. Z–15.;	精车 φ26 mm 外圆;
N190 G01 X30. Z–15.;	精车 φ30 mm 外圆起点;
N200 G01 X30. Z–30.;	精车 φ30 mm 外圆终点;
N210 G01 X32. Z–30.;	X 向退刀;
N220 G00 X100. Z100.;	返回换刀点;
N230 M05;	主轴停;
N240 M30;	程序结束, 返回程序头

表 2-3-5 为加工螺柱零件右端的程序。

表 2-3-5　加工螺柱零件右端的程序

程序内容	程序说明
O2222;	程序号;
N010 G99 M03 S600 T0101;	进给速度 mm/r, 主轴正转速度 600 r/min, 选 1 号刀;
N020 G00 X100.Z100.;	换刀点;
N030 G00 X32. Z2.;	接近工件点;
N040 G01 X30. Z–29. F0.2;	粗车 φ24 mm 外圆第一刀;
N050 G01 X32. Z–29.;	X 向退刀;
N060 G00 X32.Z2.;	Z 向退刀;
N070 G00 X28. Z2.;	粗车 φ24 mm 外圆第二刀起点;
N080 G01 X28. Z–29. F0.2;	粗车 φ24 mm 外圆第二刀终点;
N090 G01 X32. Z–29.;	X 向退刀;
N100 G00 X32. Z2.;	Z 向退刀;
N110 G00 X26. Z2.;	粗车 φ24 mm 外圆第三刀起点;
N120 G01 X26. Z–29. F0.2;	粗车 φ24 mm 外圆第三刀终点;
N130 G01 X32. Z–29.;	X 向退刀;
N140 G00 X32. Z2.;	Z 向退刀;
N150 G01 X24.5 Z2.;	粗车 φ24 mm 外圆第四刀起点;
N160 G01 X24.5 Z–29. F0.2;	粗车 φ24 mm 外圆第四刀终点;

续表

程序内容	程序说明
N170 G01 X32. Z-29.;	X向退刀;
N180 G01 X32. Z-2.;	Z向退刀;
N190 G01 X21. Z2.;	接近工件点;
N200 G01 X21. Z0. F0.1;	倒角起点;
N210 G01 X23.7. Z-1.5;	倒角终点;
N220 G01 X23.7. Z-29.;	精车φ24 mm 外圆起点;
N230 G01 X32. Z-29.;	精车φ24 mm 外圆终点;
N240 G00 X100. Z100.;	返回换刀点;
N250 T0202;	换刀宽 5 mm 的切槽刀;
N260 G00 X32. Z2.;	接近工件点;
N270 G00 X32. Z-29.;	切槽起点;
N280 G01 X20. Z-29. F0.1;	切槽终点;
N290 G04 X2.;	暂停 2 s;
N300 G01 X21. Z-29.;	退刀;
N310 G01 X24. Z-27.5;	倒角起点;
N320 G00 X32. Z-27.5;	倒角终点;
N330 G00 X100. Z100.;	返回换刀点;
N340 T0303;	换三角形外螺纹刀;
N350 G00 X26.Z2.;	循环起点;
N360 G92 X22.9 Z-27. F1.5;	车螺纹第一刀;
N370　　　 X22.3;	车螺纹第二刀;
N380　　　 X22.15;	车螺纹第三刀;
N390　　　 X22.05;	车螺纹第四刀;
N400　　　 X22.05;	精车螺纹;
N410 G00 X100. Z100.;	返回换刀点;
N420 M05;	主轴停;
N430 M30;	程序结束，返回程序头

8. 零件加工

（1）教师演示加工整个工件的准备工作及加工过程。

（2）教师讲解、演示保证尺寸公差的方法。

（3）学生分组完成工件的加工任务，教师巡回指导，及时发现并纠正学生加工中出现的问题。

（4）操作注意事项如下：

①为了保证加工基准的一致性，在多把刀具对刀时，可以先用一把刀具加工出一个基准，其他各把刀具依次以其为基准进行对刀；

②加工螺纹时主轴转速、"倍率"不能改变，否则会出现乱扣现象。

四、任务评价

按表 2-3-6 所示各项内容进行任务评价。

表 2-3-6　螺柱评分标准

	考核项目	考核要求	配分	评分标准	检测结果	得分	备注
1	外圆	$\phi30_{-0.03}^{0}$ mm	10	每超差 0.01 mm 扣 5 分			
2		$\phi30_{-0.03}^{0}$ mm	10	每超差 0.01 mm 扣 5 分			
3	长度	15 mm、29 mm、59 mm	5×3	超差全扣			
4	螺纹	大径	5	超差全扣			
5		中径	10	超差全扣			
6		两侧表面结构参数值	10	超差、降级无分			
7		牙型角	10	样板检查，超差全扣			
8		两侧表面结构参数值	10	超差全扣			
9		C1.5（两处）	2×2	超差全扣			
10	倒角	C1	4	超差全扣			
11	表面结构参数值	Ra1.6 μm（二处）	6×2	降一级扣 2 分			
12	工艺、程序	遵守工艺与程序的有关规定		违反规定扣总分 1~5 分			
13	规范操作	遵守数控车床规范操作的有关规定		违反规定扣总分 1~5 分			
14	安全文明生产	遵守安全文明生产的有关规定		违反规定扣总分 1~5 分			

学生任务实施过程的小结及反馈：

教师点评：

项目三

套类工件编程与加工

【项目描述】

学习本模块内容应掌握套类零件的安装方法及刀具安装方法；会填写加工刀具卡和工艺卡；能够完成套类零件的编程与加工，以及套类零件的检验方法；并且能理论联系实际，在编程教室利用数控仿真软件进行仿真加工，进入实训车间现场进行机床加工，分小组协作完成或单独完成加工任务。

【学习目标】

（1）能够对简单套类零件进行数控车削工艺分析。

（2）掌握镗孔刀的安装、使用与对刀方法。

（3）掌握内孔加工的程序编制方法，能够完成简单套类零件的加工。

（4）能编写简单套类零件的加工程序及加工工艺卡。

任务一　　内螺纹零件编程与加工

一、图样与技术要求

内螺纹的作用主要是与外螺纹进行配合起到连接、传递动力等作用。常见的内螺纹有粗牙三角螺纹、细牙三角螺纹、梯形螺纹、内锥螺纹等。

加工内螺纹与加工外螺纹相似，对于细牙三角螺纹可采用 G32 指令、G92 指令进行编程加工。

二、图纸分析与工艺安排

1. 图纸分析

图 3-1-1 所示为一个内螺纹零件，工件长度为 50 mm，外圆三个台阶尺寸分别为 $\phi42$ mm、$\phi36$ mm、$\phi40$ mm。内孔台阶尺寸为 $\phi22$ mm，内螺纹为 M24×1.5，退刀槽尺寸 4 mm×2 mm，技术要求为锐角倒钝 C1。

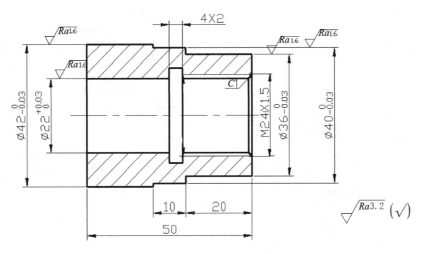

图 3-1-1　内螺纹零件

2. 工艺安排

1）确定工件的装夹方案

此零件需经二次装夹才能完成加工，第一次夹左端面，车右端，完成钻通孔、$\phi36$ mm、$\phi40$ mm 外圆，$\phi22$ mm 内孔，4 mm×2 mm 内沟槽及 M24×1.5 内螺纹的加工，第二次以 $\phi36$ mm 精车外圆为定位基准，进行 $\phi42$ mm 外圆的加工。

2）确定加工路线

（1）平端面，钻毛坯孔 $\phi20$ mm。

（2）粗、精车 $\phi36$ mm、$\phi40$ mm 外圆。

（3）粗、精车 $\phi22$ mm、$\phi22.5$ mm 内孔。

（4）加工 4 mm×2 mm 内沟槽。

（5）加工 M24×1.5 内螺纹。

（6）工件调头，夹 $\phi36$ mm 外圆。

（7）粗、精车 $\phi42$ mm 外圆。

3）填写加工刀具卡和工艺卡

加工刀具卡和工艺卡，见表 3-1-1。

表 3-1-1　加工刀具卡和工艺卡

零件图号		3-1-1	数控车床加工工艺卡	机床型号	CKA6150
零件名称		内螺纹零件		机床编号	
刀具表				量具表	
刀具号	刀具补偿号	刀具名称	刀具参数	量具名称	规格 /mm
T01	01	93°外圆精车刀	D 型刀片	游标卡尺 千分尺	0~150/0.01 25~50/0.01
T02	02	镗孔车刀	T 型刀片	内径百分表	18~35/0.01
T03	03	内切槽刀	刃宽 4 mm		

续表

工序	工艺内容	切削用量			加工性质
T04	04	内螺纹刀（图 3-1-2）		塞规	M24×1.5
		钻头 φ20		游标卡尺	0~150/0.02
工序	工艺内容	S/（r/min）	F/（mm/r）	a_p/mm	加工性质
数控车	车端面确定基准	500	—	1	手动
1	钻孔	300	—	—	手动
2	加工 φ36 mm、φ40 mm 外圆	800~1 000	0.1~0.2	0.5~3	自动
3	加工 φ22 mm、φ22.5 mm 内孔	600~800	0.05~0.1	0.3~1	自动
4	加工 4×2 内沟槽	300	0.1	4	自动
5	加工 M24×1.5 内螺纹	300	1.5	0.05~0.4	自动
数控车	调头夹 φ36 外圆，保总长	—	—	—	手动
1	加工 φ42 mm 外圆	800~1 000	0.1~0.2	0.1~1	自动

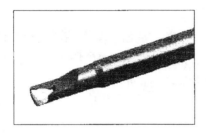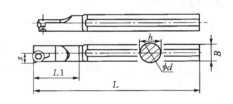

图 3-1-2　内螺纹刀 T04

三、程序编制及加工

1. 套类零件的装夹方案

套类零件的内、外圆，端面与基准轴线都有一定的形位精度要求，套类零件精加工基准可以选择外圆，但常以中心孔及一个端面为精加工基准。对不同结构的套类零件，不可能用一种工艺方案就可以保证其形位精度要求。

根据套类零件的结构特点，数控车削加工中可采用三爪卡盘、四爪卡盘或花盘装夹，由于三爪卡盘定心精度存在误差，不适于同轴度要求高的工件的二次装夹。对于能一次加工完成内外圆端面、倒角、切断的套类零件，可采用三爪卡盘装夹。较大零件经常采用四爪卡盘或花盘装夹，对于精加工零件一般可采用软卡爪装夹，对于较复杂的套类零件有时也采用专用夹具来装夹。

2. 刀具的选择

加工套类零件外圆柱面的刀具选择与轴类零件相同。加工内孔是套类零件加工的特征之一，根据内孔工艺要求，加工方法较多，常用的有钻孔、扩孔、铰孔、镗孔、磨孔、拉孔、研磨孔等，根据不同的加工方法选择适用的加工刀具。

套类零件一般包括外圆、锥面、圆弧、槽、孔、螺纹等结构。根据加工需要，常用的刀具有粗车镗孔车刀、精车镗孔车刀、内槽车刀、内螺纹车刀以及中心钻和麻花钻等。

3. 切削液的选择

在数控车削加工中，套类零件比轴类零件有更大的难度，套类零件的特性使得切削液不易达到切削区域，切削时的温度较高，切削车刀的磨损也比较严重。为了减少工件加工变形，提高加工精度，应根据不同的工件材料，选择适合的切削液，并适时调整切削液的浇注位置。

4. 调头加工时确保总体长度的方法

在前面项目中进行了半轴零件的加工，本项目开始要进行调头加工，这就需要确保零件的总体长度。常用的保总长方法是在加工前，将毛坯的两端都进行平端面操作，在平端面的过程中将毛坯加工到所要求的长度。在调头加工，进行 Z 轴对刀时使车刀与端面轻微接触，然后在对刀操作界面中输入试切削值，而不再进行平端面加工操作。

5. 加工台阶孔类零件编程举例

加工图 3-1-3 所示台阶孔类零件，加工程序见表 3-1-2、表 3-1-3。

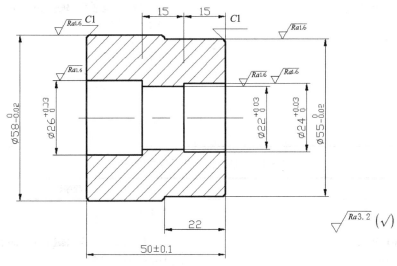

图 3-1-3　台阶孔零件

表 3-1-2　加工台阶孔零件左端程序

程序内容	程序说明
O3101;	程序号;
N1;	第 1 程序段号;
G99 M03 S700 T0202;	选 2 号刀, 主轴正转, 转速 700 r/min;
G00 X100.0 Z100.0;	快速运动到安全点;
G00 X18.0 Z2.0;	快速运动到循环点;
M08;	冷却液开;
G71 U1.0 R0.5;	粗加工 $\phi 26$ mm、$\phi 22$ mm 内孔循环;
G71 P10 Q20 U−0.5 W0.05 F0.1;	
N10 G00 G41 X27.0;	循环加工起始段程序, 刀具右补偿;
G01 Z0;	
X26.0 Z−0.5;	
Z−20.0;	
X22.0 C0.5;	
Z−36.0;	车削长度加长 1 mm;
N20 G00 G40 X18.0;	循环加工终点段程序, 取消刀具补偿;
G00 Z100.0;	快速运动到安全点;
X100.0;	
M09;	冷却液关;
M00;	程序暂停;
N2;	第 2 程序段号;
G99 M03 S800 T0202;	选 2 号刀, 主轴正转, 转速 800 r/min;
G00 X100.0 Z100.0;	快速运动到安全点;
G00 X18.0 Z2.0;	快速运动到循环点;
M08;	冷却液开;
G70 P10 Q20 F0.05;	精加工 $\phi 26$ mm、$\phi 24$ mm 内孔循环;
G00 Z100.0;	快速运动到安全点;
X100.0;	
M09;	冷却液关;
M30;	程序结束, 返回程序头

表 3-1-3　加工台阶孔零件右端程序

程序内容	程序说明
O3102;	程序号;
N1;	第 1 程序段号;
G99 M03 S600 T0202;	选 2 号刀, 主轴正转, 转速 600 r/min;
G00 X100.0 Z100.0;	快速运动到安全点;
G00 X18.0 Z2.0;	快速运动到循环点;
M08;	冷却液开;
G71 U1.0 R0.5;	粗加工 $\phi 24$ mm 内孔循环;
G71 P10 Q20 U−0.5 W0.05 F0.1;	
N10 G00 G41 X25.0;	循环加工起始段程序, 刀具右补偿;
G01 Z0;	
X24.0 Z−0.5;	
Z−15.0;	
X21.0;	
X22.0 Z−15.5;	

续表

程序内容	程序说明
N20 G00 G40 X18.0;	循环加工终点段程序，取消刀具补偿；
G00 Z100.0;	快速运动到安全点；
X100.0;	
M09;	冷却液关；
M00;	程序暂停；
N2;	第2程序段号；
G99 M03 S800 T0202;	选2号刀，主轴正转，转速800 r/min；
G00 X100.0 Z100.0;	快速运动到安全点；
G00 X18.0 Z2.0;	快速运动到循环点；
M08;	冷却液开；
G70 P10 Q20 F0.05;	精加工 ϕ30 mm 内孔循环；
G00 Z100.0;	快速运动到安全点；
X100.0;	
M09;	冷却液关；
M30;	程序结束，返回程序头

6. 加工内沟槽类零件编程举例

1）刀具

加工内沟槽类零件常用刀具如图 3-1-4 所示。

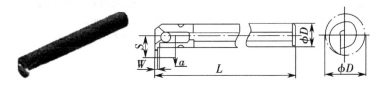

图 3-1-4 内沟（切）槽刀

2）编程举例

图 3-1-5 所示为一套类带内沟槽零件，工件长度为 50 mm，外圆三个台阶尺寸分别为 ϕ42 mm、ϕ36 mm、ϕ40 mm。内孔三个台阶尺寸分别为 ϕ24 mm、ϕ22 mm、ϕ24 mm，包含有 4 mm×2 mm、10 mm×3 mm 两个内沟槽，技术要求为锐角倒钝 C0.5。加工程序见表 3-1-4、表 3-1-5。

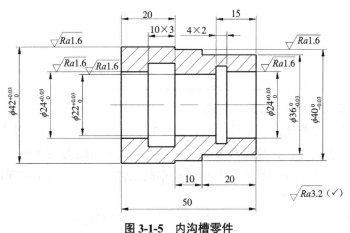

图 3-1-5 内沟槽零件

表 3-1-4　加工内沟槽零件右端程序

程序内容	程序说明
O3301;	程序号;
N1;	第 1 程序段号;
G99 M03 S600 T0303;	选 3 号刀，主轴正转，转速 600 r/min;
G00 X100.0 Z100.0;	快速运动到安全点;
G00 X20.0 Z2.0;	快速运动到循环点;
Z−15.0;	
M08;	冷却液开;
G01 X28.0;	加工 4 mm×2 mm 内沟槽;
X20.0;	退刀;
G00 Z100.0;	快速运动到安全点;
X100.0;	
M09;	冷却液关;
M05;	主轴停转;
M30;	程序结束，返回程序头

表 3-1-5　加工内沟槽零件左端程序

程序内容	程序说明
O3302;	程序号;
N1;	第 1 程序段号;
G99 M03 S600 T0303;	选 3 号刀，主轴正转，转速 600 r/min;
G00 X100.0 Z100.0;	快速运动到安全点;
G00 X20.0 Z2.0;	快速运动到循环点;
Z−20.0;	
M08;	冷却液开;
G01 X30.0;	加工 10 mm×3 mm 内沟槽;
X22.0;	
G00 W4.0;	
G01 X30.0;	
X22.0;	
G00 W2.0;	
G01 X30.0;	
Z−20.0;	
X22.0;	退刀;
G00 Z100.0;	快速运动到安全点;
X100.0;	
M09;	冷却液关;
M05;	主轴停转;
M30;	程序结束，返回程序头

7. 内螺纹车刀的选择、刃磨和装夹

（1）内螺纹车刀的选择。内螺纹车刀是根据车削方法和工件材料及形状来选择的。车刀的尺寸大小受到螺纹孔径尺寸限制，一般内螺纹车刀的刀头径向长度应比孔径小 3~5 mm，否则退刀时易碰伤牙顶，甚至不能车削工件。刀杆的大小在保证排屑的前提下，应尽量选择较粗的刀杆。

（2）车刀的刃磨和装夹。内螺纹车刀的刃磨方法和外螺纹车刀的基本相同，但是刃磨刀尖时要注意它的平分线必须与刀杆垂直，否则车内螺纹时会出现刀杆碰伤内孔的现象，刀尖宽度应符合要求，一般为 0.1 mm 乘以螺距。

8. 三角形内螺纹尺寸的计算

以 M24×1.5 内螺纹为例，计算螺纹尺寸。

（1）螺纹大径 $D=$ 公称直径 $=24$ mm

（2）螺纹孔径 $D_{孔}=$ 公称直径 $D-1.082\ 5\times$ 螺距 $P=24-1.082\ 5\times1.5=22.376$ mm

9. 内螺纹加工的注意事项

（1）车内螺纹前，先把工件的内孔、端面及倒角车好。

（2）进刀切削方式和外螺纹相同，螺距小于 1.5 mm 或铸铁螺纹采用直进法；螺距大于 2 mm 采用左右切削法。车内螺纹时目测困难，一般根据观察到的排屑情况进行左右分刀切削，并判断螺纹的表面质量。

10. 零件加工程序

根据图 3-1-1 所示零件，确定工件加工装夹方案、加工路线及采用的刀具和切削用量，根据工艺过程将工序内容划分为五个部分，并对应编制五个程序完成加工，在这里只列出加工内螺纹零件内孔程序（表 3-1-6）和加工内螺纹零件右端程序（表 3–1–7）。

表 3-1-6　加工内螺纹零件内孔程序

程序内容	程序说明
O3009;	程序号;
N1;	第 1 程序段号;
G99 M03 S600 T0202;	选 2 号刀，主轴正转，转速 600 r/min;
G00 X100.0 Z100.0;	快速运动到安全点;
G00 X19.0 Z2.0;	快速运动到循环点;
M08;	冷却液开;
G71 U1.0 R0.5;	粗加工内孔循环;
G71 P10 Q20 U–0.5 W0.05 F0.1;	
N10 G00 G41 X24.0;	循环加工起始段程序，刀具右补偿;
G01 Z0;	
X22.38 Z–1.0;	
Z–25.0;	
X22.0;	
Z–51.0;	
N20 G00 G40 X19.0;	循环加工终点段程序，取消刀具补偿;
G00 Z100.0;	快速运动到安全点;
X100.0;	
M09;	冷却液关;
M00;	程序暂停;
N2;	第 2 程序段号;
G99 M03 S800 T0202;	选 2 号刀，主轴正转，转速 800 r/min;
G00 X100.0 Z100.0;	快速运动到安全点;
G00 X19.0 Z2.0;	快速运动到循环点;
M08;	冷却液开;
G70 P10 Q20 F0.05;	精加工内孔循环;

续表

程序内容	程序说明
G00 Z100.0;	快速运动到安全点;
X100.0;	
M09;	冷却液关
M30;	程序结束，返回程序头

表 3-1-7　加工内螺纹零件右端程序

程序内容	程序说明
O3010;	程序号;
N1;	第1程序段号;
G99 M03 S300 T0404;	选4号刀，主轴正转，转速300 r/min;
G00 X100.0 Z100.0;	快速运动到安全点;
G00 X20.0 Z5.0;	快速运动到循环点;
M08;	冷却液开;
G92 X22.8.0 Z-22.5 F1.5;	螺纹加工循环开始;
X23.2;	
X23.6;	
X23.9;	
X23.95;	
X24.0;	
X24.0;	
G00 Z100.0;	快速移动到安全点;
X100.0;	
M05;	主轴停止;
M09;	冷却液关;
M30;	程序结束，返回程序头

11. 零件加工

（1）教师演示加工整个工件的准备工作及加工过程。

（2）教师讲解、演示保证尺寸公差的方法。

（3）学生分组完成工件的加工任务，教师巡回指导，及时发现并纠正学生加工中出现的问题。

（4）操作注意事项。

①加工内螺纹时应注意起刀点不能与工件端面距离过近，以免左侧刀体与工件发生碰撞。

②加工内螺纹时每刀的切削量要逐刀减小，这一点与加工外螺纹类似，并且由于内螺纹刀刀体较细，伸出长度较长，因此切削量不宜过大。

③选择内螺纹车刀时要注意刀体直径，不要使车刀体与工件发生干涉。

12. 零件测量

内螺纹的检验方法有两种：综合检验和单项检验，通常我们进行综合检验。

如图 3-1-6 所示，内螺纹塞规分为通端（T）与止端（Z），如果被测内螺纹能够与塞规通端旋合通过，且与塞规止端不完全旋合通过（螺纹塞规只允许与被测螺纹两段旋合，旋合量不得超过两个螺距），表明被测内螺纹的中径没有超过其最大实体牙型的中径，且单一中径没有超出其最小实体牙型的中径，那么就可以保证旋合性和连接强度，则被测内螺纹中径合格，否则不合格。

图 3-1-6　内螺纹塞规

四、任务评价

按表 3-1-8 所示各项内容进行任务评价。

表 3-1-8　内螺纹零件评分标准

	考核项目	考核要求	配分	评分标准	检测结果	得分	备注
1	外圆	$\phi42_{-0.03}^{0}$ mm	12	超差 0.01 mm 扣 6 分			
		$\phi40_{-0.03}^{0}$ mm	12	超差 0.01 mm 扣 6 分			
		$\phi36_{-0.03}^{0}$ mm	12	超差 0.01 mm 扣 6 分			
2	内孔	$\phi22_{0}^{0.03}$ mm	15	超差 0.01 mm 扣 7 分			
3	内槽	4 mm×2 mm	8	不合格无分			
4	内螺纹	M24×1.5	10	不合格无分			
		21 mm	8	不合格无分			
	长度	50±0.1 mm	7	超差 0.1 扣 4 分			
	表面结构参数值	Ra3.2 μm（四处）	16	Ra 值大 1 级无分			
5	工艺、程序	遵守工艺与程序的有关规定		违反规定扣总分 1~5 分			
6	规范操作	遵守数控车床规范操作的有关规定		违反规定扣总分 1~5 分			
7	安全文明生产	遵守安全文明生产的有关规定		违反规定扣总分 1~5 分			
学生任务实施过程的小结及反馈：							
教师点评：							

任务二　　轴承套类零件编程与加工

一、图样与技术要求

图 3-2-1 所示为一轴承套零件，工件长度为 60 mm，外圆台阶尺寸分别为 ϕ58 mm、ϕ45 mm，内孔台阶尺寸为 ϕ30 mm，内槽尺寸 32 mm×20 mm，技术要求为锐角倒钝 $C1$、$C2$。

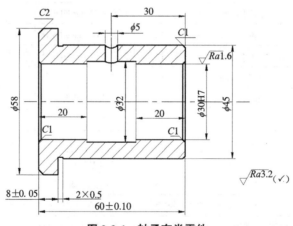

图 3-2-1　轴承套类零件

二、图纸分析与工艺安排

1. 图纸分析

图 3-2-1 为一简单轴承套类零件，该零件表面由两个台阶组成，零件图尺寸标注完整，符合数控加工尺寸标注要求；轮廓描述清楚完整；零件材料可选铜料，加工切削性能较好，无热处理和硬度要求。

2. 工艺安排

1）确定工件的装夹方案

此零件为一通孔套类零件，尺寸要求较高，应使用 ϕ60 mm 以上毛坯，长度

65 mm。首先加工 $\phi58$ mm 和 $\phi45$ mm 外圆直径，再加工 $\phi30$ mm 内孔和 $\phi32$ mm 内沟槽。掉头并找正装夹 $\phi45$ mm 外圆完成倒角及工件全长的加工。

2）确定加工路线

（1）平端面，钻毛坯孔 $\phi28$ mm。

（2）粗、精车 $\phi58$ mm、$\phi45$ mm 外圆。

（3）粗、精车 $\phi30$ mm 内孔。

（4）加工 $\phi32$ mm 内沟槽。

（5）工件调头并找正，夹 $\phi45$ mm 外圆。

（6）加工工件全长尺寸 60 mm 并倒角。

3）填写加工刀具卡和工艺卡

加工刀具卡和工艺卡，见表 3-2-1。

表 3-2-1 加工刀具卡和工艺卡

零件图号	3—2—1	数控车床加工工艺卡		机床型号	CKA6150
零件名称	轴承套类零件			机床编号	
刀具表				量具表	
刀具号	刀具补偿号	刀具名称	刀具参数	量具名称	规格 /mm
T01	01	93°外圆精车刀	D 型刀片	游标卡尺	0~150/0.01
T02	02	镗孔车刀	T 型刀片	内径百分表	18~35/0.01
T03	03	内切槽刀	刃宽 4 mm		
		钻头 $\phi28$		游标卡尺	0~150/0.02

工序	工艺内容	切削用量			加工性质
		$S/$（r/min）	$F/$（mm/r）	$a_p/$mm	
数控车	车外圆、平端面确定基准	500	—	1	手动
1	钻孔	300			手动
2	加工 $\phi58$ mm、$\phi45$ mm 外圆	800~1 000	0.1~0.2	0.5~3	自动
3	加工 $\phi30$ mm 内孔	600~800	0.05~0.1	0.3~1	自动
4	加工 $\phi32$ mm 内沟槽	600	0.1	4	自动
数控车	调头夹 $\phi45$ mm 外圆	—			手动
1	加工工件全长 60 mm	800~1 000	0.1~0.2	0.5~3	自动

三、程序编制及加工

1. 滚动轴承结构

常见的滚动轴承一般由两个套圈（即内圈、外圈）、滚动体和保持架等基本元件组成（图 3-2-2）。通常内圈与轴颈相配合且随轴一起转动，外圈装在机架的轴承座孔内固定不动。当内、外圈相对旋转时，滚动体在内、外圈的滚道上滚动，保持架使滚动体均匀分布并避免相邻滚动体之间的接触摩擦和磨损。

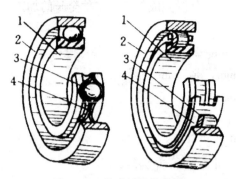

图 3-2-2　滚动轴承的结构

1—内圈；2—外圈；3—滚动体；4—保持架

滚动轴承的内、外圈和滚动体一般采用专用的滚动轴承钢制造，如 GCr9、GCr15、GCr15SiMn 等，保持架则常用较软的材料如低碳钢板经冲压制成，或用铜合金、塑料等制成。

2. 滚动轴承的 4 个基本参数

1）接触角

如图 3-2-3 所示，滚动轴承中滚动体与外圈接触处的法线和垂直于轴承轴心线的平面的夹角 α 称为接触角。接触角越大，轴承承受轴向载荷的能力越大。

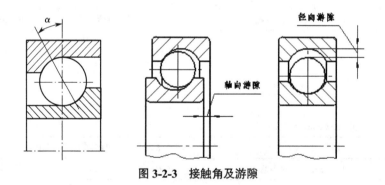

图 3-2-3　接触角及游隙

2）游隙

滚动体与内、外圈滚道之间的最大间隙称为轴承的游隙。

3）偏位角

如图 3-2-4 所示，轴承内、外圈轴线相对倾斜时所夹锐角，称为偏位角。能自动适应偏位角的轴承，称为调心轴承。

4）极限转速

滚动轴承在一定的载荷和润滑的条件下，允许的最高转速称为极限转速，其具体数值见有关手册。

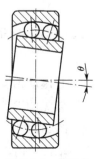

图 3-2-4　偏位角

3. 滚动轴承的类型

滚动轴承的类型很多，下面介绍几种常见的分类方法。

（1）按滚动体的形状分，可分为球轴承和滚子轴承两大类。

（2）按滚动体的列数，滚动轴承又可分为单列、双列及多列滚动轴承。

（3）按工作时能否调心可分为调心轴承和非调心轴承，调心轴承允许的偏位角大。

（4）按承受载荷方向不同，可分为向心轴承和推力轴承两类。

①向心轴承主要承受径向载荷，公称接触角 $\alpha=0°$ 的轴承称为径向接触轴承；$0°<\alpha\leqslant45°$ 的轴承，称为角接触向心轴承。

②推力轴承主要承受轴向载荷，公称接触角 $45°<\alpha<90°$ 的轴承，称为角接触推力轴承，其中 $\alpha=90°$ 的称为轴向接触轴承，也称推力轴承。接触角越大，承受径向载荷的能力越小，承受轴向载荷的能力越大，轴向推力轴承只能承受轴向载荷。

4. 零件加工程序

根据图 3-2-1 所示零件，分析工件加工要求，确定加工装夹方案、加工路线及采用的刀具和切削用量，根据工艺过程将工序内容划分为三个部分，并对应编制三个程序，在这里只列出车内孔和内沟槽的程序。表 3-2-2 为加工轴承套类零件内孔的程序，表 3-2-3 为加工轴承套类零件内沟槽的程序。

表 3-2-2　加工轴承套类零件内孔的程序

程序内容	程序说明
O3501;	程序号;
N1;	第 1 程序段号;
G99 M03 S600 T0202;	选 2 号刀，主轴正转，转速 600 r/min;
G00 X100.0 Z100.0;	快速运动到安全点;
G00 X25.0 Z2.0;	快速运动到循环点;
M08;	冷却液开;
G71 U1.0 R0.5;	粗加工内孔循环;
G71 P10 Q20 U−0.5 W0.05 F0.1;	
N10 G00 G41 X28.0;	循环加工起始段程序，刀具右补偿;
G01 Z0;	
X30 Z−1.0;	
Z−61;	
N20 G00 G40 X25.0;	循环加工终点段程序，取消刀具补偿;
G00 Z100.0;	快速运动到安全点;
X100.0;	
M09;	冷却液关;
M00;	程序暂停;
N2;	第 2 程序段号;
G99 M03 S800 T0202;	选 2 号刀，主轴正转，转速 800 r/min;
G00 X100.0 Z100.0;	快速运动到安全点;
G00 X25.0 Z2.0;	快速运动到循环点;
M08;	冷却液开;
G70 P10 Q20 F0.05;	精加工内孔循环;
G00 Z100.0;	快速运动到安全点;
X100.0;	

续表

程序内容	程序说明
M09; M30;	冷却液关; 程序结束,返回程序头

表 3-2-3　加工轴承套类零件内沟槽的程序

程序内容	程序说明
O3502;	程序号;
N1;	第 1 程序段号;
G99 M03 S600 T0303;	选 4 号刀,主轴正转,转速 600 r/min;
G00 X100.0 Z100.0;	快速运动到安全点;
X25.0	
G00 X25.0 Z-24.2;（+刀宽）	快速运动到循环点（注:加上切刀宽度 4 mm）;
M08;	冷却液开;
G01 X31.8 Z-24.2 F0.05;	加工开始,第一刀;
G00 X25;	
Z-28.2;	第二刀;
G01 X31.8;	
G00 X25;	
Z-32.2;	第三刀;
G01 X31.8;	
G00 X25;	
Z-35.8;	第四刀;
G00 X25;	
Z-36;	精车宽槽;
G01 X32;	
Z-24;	
X25;	
G00 Z100.0;	快速移动到安全点;
X100.0;	
M05;	主轴停止;
M09;	冷却液关;
M30;	程序结束,返回程序头

5. 零件加工

（1）教师演示加工整个工件的准备工作及加工过程。

（2）教师讲解、演示保证尺寸公差的方法。

（3）学生分组完成工件的加工任务,教师巡回指导,及时发现并纠正学生加工中出现的问题。

（4）进行内槽加工时切削力较大,要时刻注意排屑以及冷却液的浇注情况,一旦出现异常要及时解决。

四、零件测量

1. 内测千分尺

内测千分尺如图 3-2-5 所示，是测量小尺寸内径和内侧面槽宽度的测量工具，其特点是易找正内孔直径，测量方便。国产的内测千分尺的读数值为 0.01 mm，常见测量范围有 5~30 mm 和 25~50 mm 两种，图 3-2-5 所示的是 5~30 mm 的内测千分尺。内测千分尺的读数方法与外径千分尺相同，只是套筒上的刻线尺寸与外径千分尺相反，另外它的测量方向和读数方向也都与外径千分尺相反。

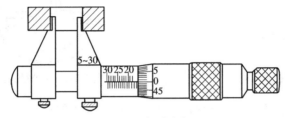

图 3-2-5　内测千分尺

2. 三爪内径千分尺

三爪内径千分尺，适用于测量中小直径的精密内孔，尤其适用于测量深孔的直径。测量范围如下：6~8 mm，8~10 mm，10~12 mm，11~14 mm，14~17 mm，17~20 mm，20~25 mm，25~30 mm，30~35 mm，35~40 mm，40~50 mm，50~60 mm，60~70 mm，70~80 mm，80~90 mm，90~100 mm。三爪内径千分尺的零位，必须在标准孔内进行校对。

三爪内径千分尺的工作原理：图 3-2-6 是测量范围为 11~14 mm 的三爪内径千分尺，当顺时针旋转测力装置 6 时，测微螺杆 3 会被旋转，并沿着螺纹轴套 4 的螺旋线方向移动，于是测微螺杆端部的方形圆锥螺纹就推动三个测量爪 1 作径向移动。扭簧 2 的弹力使测量爪紧紧地贴合在方形圆锥螺纹上，并随着测微螺杆的进退而伸缩。

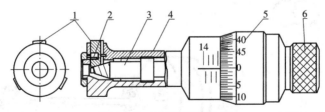

图 3-2-6　三爪内径千分尺

1—测量爪；2—扭簧；3—测微螺杆；4—螺纹轴套；5—微分筒；6—测力装置

三爪内径千分尺的方形圆锥螺纹的径向螺距为 0.25 mm，即当测力装置顺时针旋转一周时测量爪 1 就向外移动（半径方向）0.25 mm，三个测量爪组成的圆周直径就要增加 0.5 mm，即微分筒旋转一周时，测量直径增大 0.5 mm，而微分筒的圆周上刻着 100个等分格，所以它的读数值为 0.5 mm÷100=0.005 mm。

3. 内径百分表的使用方法

内径百分表用来测量圆柱孔，它附有成套的可调测量头，使用前必须先进行组合和校对零位，如图 3-2-7 所示。

图 3-2-7　内径百分表

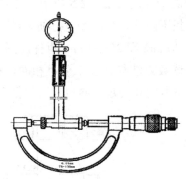

图 3-2-8　用外径百分尺调整尺寸

组合时，将百分表装入连杆内，使小指针指在 0~1 的位置上，长针和连杆轴线重合，刻度盘上的字应垂直向下，以便于测量时观察，装好后应予紧固。粗加工时，最好先用游标卡尺或内卡钳测量。因内径百分表同其他精密量具一样属于贵重仪器，其好坏与精确程度直接影响到工件的加工精度和其使用寿命，且粗加工时工件加工表面粗糙不平易引起测量不准确。因此，须对内径百分表加以爱护和保养，精加工时再用其进行测量。

测量前应根据被测孔径大小用外径百分尺调整好尺寸后才能使用，如图 3-2-8 所示。在调整尺寸时，正确选用可换测量头的长度及其伸出距离，应使被测尺寸在活动测量头的总体移动量中间位置。

测量时，连杆中心线应与工件中心线平行，不得歪斜，同时应在圆周上多测几个点，找出工件孔径的实际尺寸，看是否在公差范围内，如图 3-2-9 所示。

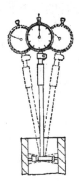
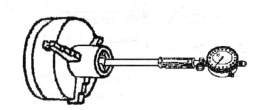

图 3-2-9　内径百分表的使用方法

五、任务评价

按表 3-2-4 所示各项内容进行任务评价。

表 3-2-4 轴承套类零件评分标准

	考核项目	考核要求	配分	评分标准	检测结果	得分	备注
1	内轮廓	ϕ45 mm/Ra1.6 μm 三处	40	超差 0.01 mm 扣 2 分 超差 0.02 mm 不得分			
		ϕ53 mm	5	超差不得分			
		ϕ30 H7 mm/Ra1.6 μm	25	超差 0.01 mm 扣 2 分 超差 0.02 mm 不得分			
		ϕ32 mm×20	6	超差不得分			
		2×0.5 mm	4	超差不得分			
		60±0.10 mm	6	超差不得分			
		8±0.05 mm	6	超差不得分			
		C1、C2，四处	2×4	不合格不得分			
2	工艺、程序	遵守工艺与程序的有关规定		违反规定扣总分 1~5 分			
3	规范操作	遵守数控车床规范操作的有关规定		违反规定扣总分 1~5 分			
4	安全文明生产	遵守安全文明生产的有关规定		违反规定扣总分 1~5 分			

学生任务实施过程的小结及反馈：

教师点评：

项目四

复杂零件编程与加工

【项目描述】

学习本项目内容应掌握典型轴套配合零件加工程序编制和加工刀具的选择方法；掌握配合件的加工方法、轴套配合零件的加工和检验方法；并且能理论联系实际，在编程教室利用数控仿真软件进行仿真加工，进入实训车间现场进行机床加工，分小组协作完成或单独完成加工任务。

【学习目标】

（1）掌握典型轴套配合零件加工程序编制方法。

（2）掌握配合件的加工方法。

（3）掌握轴套配合零件的加工和检验方法。

（4）能够综合运用已学专业理论知识，独立分析比较复杂零件的加工工艺过程。

（5）能够根据零件的技术要求选择正确的加工方案和合适的刀具，能够选择合适的装夹方式。

（6）掌握典型心轴类零件的加工程序编制方法，能合理选择切削用量。

（7）掌握加工中误差现象的产生原因及预防和消除方法。

任务一　　　轴套配合零件加工

一、图样与技术要求

如图 4-1-1 所示为一组轴套配合组合零件，轴与套以内外圆圆锥表面配合总长度为 78 ± 0.1 mm，轴与套以内外圆柱面配合总长度为 103 ± 0.1 mm。材料为 45 钢，毛坯为 $\phi 45$ mm×75 mm、$\phi 60$ mm×53 mm 棒料各一根。

根据图 4-1-1，可知零件技术要求如下：

（1）锐边去毛刺倒棱，未注倒角 $C1$，轴右端面允许打中心孔。

（2）配作圆锥面接触大于 70%。

（3）轴、套类面结构参数 Ra 值全部为 1.6 μm。

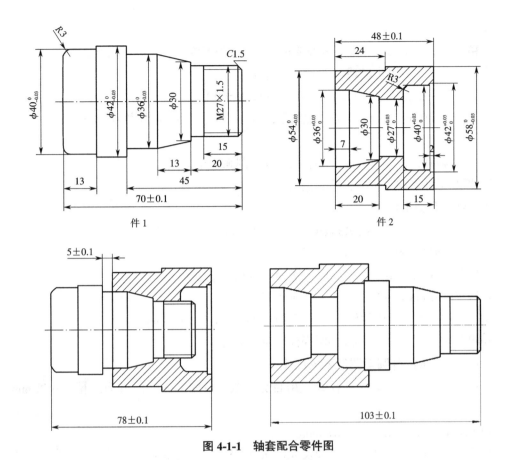

图 4-1-1　轴套配合零件图

二、图纸分析

1. 图纸分析

分析工件组合后，难保证的尺寸精度主要有：配合长度尺寸 103 ± 0.1 mm、78 ± 0.1 mm，间隙尺寸 5 ± 0.1 mm。其他难保证的组合精度有：内、外圆锥接触面积大于 70%，内、外圆弧，内、外圆柱面配合。此零件加工的难点在于保证各项配合的精度。

2. 工艺安排

1）确定工件的加工方案

分析图纸，轴、套均采用两次装夹后完成粗、精加工的加工方案。进行数控车削加工时，加工的起始点定在离工件毛坯 2 mm 的位置。应尽可能采用沿轴向切削的方式进行加工，以提高加工过程中工件与刀具的刚性。

2）确定加工路线

（1）件 1 轴加工路线。

①采用三爪自定心卡盘夹持 $\phi45$ mm 外圆，毛坯伸出卡盘 50 mm 找正夹紧，车端面。

②粗车 $\phi43$ mm、$\phi36$ mm、$\phi26.9$ mm 外圆及外锥，径向留量 0.5 mm，轴向留量

0.1 mm。

③精车ϕ43 mm、ϕ36 mm、ϕ26.9 mm 外圆及外圆锥至图纸要求尺寸。

④加工外螺纹并用止、通规检验螺纹精度。

⑤掉头装夹找正，手工车端面，保证总长 70±0.1 mm。

⑥粗车ϕ40 mm、ϕ42 mm 外圆及 R3 mm 圆弧，径向留量 0.5 mm 轴向留量 0.1 mm。

⑦精车ϕ40 mm、ϕ42 mm 外圆及 R3 mm 圆弧，至图纸要求尺寸。

⑧拆卸工件，并对工件去毛刺倒棱。

（2）件 2 套加工路线。

①采用三爪自定心卡盘夹持ϕ60 mm 外圆毛坯伸出卡盘 30 mm 找正夹紧，手工钻通孔，孔径为ϕ25 mm；车端面。

②粗车ϕ36 mm、ϕ27 mm 内圆及内锥，径向留量 0.5 mm。

③精车ϕ36 mm、ϕ27 mm 内圆至图纸尺寸。

④车内锥与外锥配作并修正内锥面保证配合间隙 5±0.1 mm。

⑤粗车ϕ59 mm、ϕ54 mm 外圆径向留量 0.5 mm。

⑥精车ϕ54 mm 外圆至图纸要求尺寸。

⑦掉头装夹于ϕ54 mm 直径处，用百分表校正ϕ59 mm 外圆表面，精车ϕ58 mm 外圆。

⑧手工车右端面保证 48±0.1 mm。

⑨粗车ϕ42 mm、ϕ40 mm 内圆及 R3 mm 圆弧，径向留量 0.5 mm。

⑩精车ϕ42 mm、ϕ40 mm 内圆及 R3 mm 圆弧至图纸尺寸，并保证配合长度 103±0.1 mm。

拆卸工件，去毛刺倒棱，检查各项加工精度。

3）填写加工刀具卡和工艺卡

加工刀具卡和工艺卡，见表 4-1-1。

表 4-1-1　加工刀具卡和工艺卡

零件图号		4-1-1	数控车床加工工艺卡	机床型号	CKA6150
零件名称		轴套配合零件		机床编号	
刀具表				量具表	
序号	刀具号	刀具补偿号	刀具名称	刀具参数	量具名称	规格 /mm
1	T01	01	90°外圆粗车刀	C 型刀片	游标卡尺 千分尺	25~50/0.01 50~75/0.01 0~150/0.02
2	T02	02	90°外圆精车刀	D 型刀片	游标卡尺 千分尺	25~50/0.01 50~75/0.01 0~150/0.02
3	T03	03	外螺纹刀	刀尖角60°	通、止规	M27×1.5

续表

| 4 | T04 | 04 | 91°镗孔车刀 | C 型刀片 | | 内径百分表 | 18~50/0.01 |
| 5 | | | 钻头 φ25 | | | 游标卡尺 | 0~150/0.02 |

| 工序 | 工艺内容 | 切削用量 | | | 加工性质 |
		S/（r/min）	F/（mm/r）	a/mm	
数控车（轴）	采用三爪自定心卡盘夹持毛坯，车端面	600~800	0.2	—	手动
1	粗、精车轴右端外轮廓	600~1200	0.1~0.2	0.25~2	自动
2	加工外螺纹	600	1.5	—	自动
数控车	掉头装夹找正，手工车端面，保证总长 70±0.1 mm	600~800	0.2	—	手动
1	粗、精车轴左端外轮廓	600~1200	0.1~0.2	0.25~2	自动
2	拆卸工件，并对工件去毛刺倒棱				
数控车（套）	采用三爪自定心卡盘夹持手工钻通孔，孔径为 φ25 mm；车端面	300~400	—	—	手动
1	粗、精车套右端内轮廓	600~1200	0.1~0.2	0.25~2	自动
2	内锥与外锥配作	1200	0.1	—	自动
3	粗、精车套左端外轮廓	600~1200	0.1~0.2	0.25~2	自动
数控车	掉头装夹 φ54 mm 外圆，用百分表校正 φ59 mm 外圆表面				
1	车端面保证 48±0.1 mm	600~800	0.2	—	手动
2	粗、精车套右端内外轮廓	600~1 200	0.1~0.2	0.25~2	自动
3	去毛刺倒棱，检查各项加工精度				

三、程序编制与零件加工

由于工件在长度方面的要求较低，根据编程原点的确定原则，工件的编程原点可取在装夹后工件的右端面与主轴轴线相交的交点上。

根据图 4-1-1 所示零件，分析工件加工路线，确定加工装夹方案及采用的刀具和切削用量，根据加工步骤将工序划分为四个部分，对应编制四个程序并完成加工。表 4-1-2 为加工轴右端的外部轮廓的程序，表 4-1-3 为加工轴左端的外部轮廓的程序，表 4-1-4 为加工套左端的内部轮廓的程序，表 4-1-5 为加工套右端的内部轮廓的程序。

表 4-1-2　加工轴右端外部轮廓的程序

程序内容	程序说明
O4101；	程序号；
N1；	第 1 程序段号；
G99 M03 S600 T0101；	选 1 号刀，主轴正转，转速 600 r/min；
G00 X100.0 Z100.0；	快速运动到安全点；
G00 G42 X45.0 Z2.0；	快速运动到循环点，填加刀尖圆弧半径右补偿；
M08；	冷却液开；
G71 U1.0 R0.5；	粗加工循环；
G71 P10 Q20 U0.5 W0.1 F0.2；	
N10 G00 X24.0；	循环加工起始程序段；
G01 Z0. F0.1；	
G01 X26.9. Z−1.5；	
G01 Z−20.；	
G01 X30.；	
G01 X36.W−13.；	
G01 Z−45.；	
G01 X43.；	
G01 W−5.；	
N20 G00 X45.；	循环加工终点程序段；
G00 X100.0 Z100.0；	快速运动到安全点；
M05；	冷却液关；
M00；	程序暂停；
N2；	第 2 程序段号；
G99 M03 S1200 T0202；	选 2 号刀，主轴正转，转速 1200 r/min；
G00 X100.0 Z100.0；	快速运动到安全点；
G00 X45.0 Z2.0；	快速运动到循环点；
G70 P10 Q20；	精加工循环；
G00 G40 X100.0 Z100.0；	快速运动到安全点，取消刀具补偿；
M05；	主轴停；
M00；	程序暂停；
N3；	第 3 程序段号；
G99 M03 S600 T0303；	选 3 号刀，主轴正转，转速 600 r/min；
G00 X28. Z5.；	快速运动到螺纹切削循环起点；
G92 X26.4 Z−15. F1.5；	螺纹切削循环；
X25.9；	
X25.5；	
X25.2；	
X25.1；	
X25.05；	
G00 X100.Z100.；	退刀；
M30；	程序结束并返回第一程序段

表 4-1-3　加工轴左端外部轮廓的程序

程序内容	程序说明
O4102；	程序号；
N1；	第 1 程序段；
G99 M03 S600 T0101；	选 1 号刀，主轴正转，转速 600 r/min；
G00 X100.0 Z100.0；	快速运动到安全点；
G00 X46.0 Z5.0；	快速运动到循环点；
M08；	冷却液开；
G71 U1.0 R0.5；	粗加工循环；
G71 P10 Q20 U0.5 W0.1 F0.2；	
N10 G00 X34.0；	循环加工起始程序段；
G01 Z0.F0.1；	
G03 X40 W-3.R3.；	
G01 Z-13.；	
G01 X42.0；	
G01 W-13.；	
N20 G00 X46.；	循环加工终点程序段；
G00 X100. Z100.；	快速运动到安全点；
M09；	冷却液关；
M05；	主轴停；
M00；	程序暂停；
N2；	第 2 程序段号；
G99 M03 S1200 T0202；	选 2 号刀，主轴正转，转速 1200 r/min；
G00 G42 X46.0 Z5.0；	快速运动到循环点，刀具右补偿；
M08；	冷却液开；
G70 P10 Q20；	精加工循环；
G00 G40 X100. Z100.；	快速运动到安全点，取消刀尖圆弧半径补偿；
M09；	冷却液关；
M30；	程序结束并返回第一程序段

表 4-1-4　加工套左端内部轮廓的程序

程序内容	程序说明
O4103；	程序号；
N1；	第 1 程序段号；
G99 M03 S600 T0404；	选 4 号刀，主轴正转，转速 600 r/min；
G00 X100.0 Z100.0；	快速运动到安全点；
G00 G41 X25.0 Z2.0；	快速运动到循环点，刀具左补偿；
M08；	冷却液开；
G71 U1.0 R0.5；	粗加工循环；
G71 P10 Q20 U-0.5 W0.05 F0.2；	
N10 G00 X36.0；	循环加工起始段程序；
G01 Z-7.F0.1；	
G01 X30.W-13.；	
G01 X27.；	
G01 W-15.；	
N20 G00 X25.；	循环加工终点段程序；
G00 Z100.；	快速运动到安全点；
G00 X100.；	

续表

程序内容	程序说明
M09;	冷却液关;
M00;	程序暂停;
N2;	第2程序段号;
G99 M03 S1200 T0404;	选4号刀，主轴正转，转速1200 r/min;
G00 X25.0 Z2.0;	快速运动到循环点;
M08;	冷却液开;
G70 P10 Q20;	精加工循环（锥体部分跳过）;
G00 G40 X100.Z100.;	快速运动到安全点;
M09;	冷却液关;
M30;	程序结束并返回第一程序段

表 4-1-5 加工套右端内部轮廓的程序

程序内容	程序说明
O4104;	程序号;
N1;	第1程序段号;
G99 M03 S600 T0404;	选4号刀，主轴正转，转速600 r/min;
G00 X100.0 Z100.0;	快速运动到安全点;
G00 G41 X25.0 Z2.0;	快速运动到循环点，刀具左补偿;
M08;	冷却液开;
G71 U1.0 R0.5;	粗加工循环;
G71 P10 Q20 U−0.5 W0.F0.2;	
N10 G00 X42.;	循环加工起始段程序;
G01 Z−2.;	
G01 X40.;	
G01 Z−12.	
G03 X34.W−3.R3.	
N20 G01 X25.;	循环加工终点段程序;
G00 Z100.;	
G00 X100.	快速运动到安全点;
M09;	冷却液关;
M00;	程序暂停;
N2;	第2程序段号;
G99 M03 S1200 T0404;	选4号刀，主轴正转，转速1200 r/min;
G00 X25.0 Z2.0;	快速运动到循环点;
M08;	冷却液开;
G70 P10 Q20;	精加工循环;
G00 Z100.;	快速运动到安全点;
G00 G40 X100.	取消刀具补偿;
M09;	冷却液关;
M30;	程序结束返回程序头

四、评分表

按表 4-1-6 所示各项内容进行任务评价。

表 4-1-6 轴套配合件评分标准

	考核项目		考核要求	配分	评分标准	检测结果	得分	备注
1	件1	外径	$\phi42_{-0.03}^{0}$ mm/Ra1.6 μm	4+1	超差全扣			
2			$\phi40_{-0.03}^{0}$ mm/Ra1.6 μm	4+1	超差全扣			
3			$\phi36_{-0.03}^{0}$ mm/Ra1.6 μm	4+1	超差全扣			
4		长度	20 mm	2	超差全扣			
5			45 mm	2	超差全扣			
6			13 mm（两处）	4	超差全扣			
7			70±0.1 mm	3	超差全扣			
8		圆弧	$R3$ mm/Ra1.6 μm	2+1	R 规检测			
9		螺纹	M27×1.5 mm/ Ra3.2 μm	6+2	螺纹通、止规测量			
10		倒角	C1.5	2	目测			
11	件2	外径	$\phi54_{-0.03}^{0}$ mm/Ra1.6 μm	4+1	超差 0.01 mm 扣 2 分			
12			$\phi58_{-0.03}^{0}$ mm/Ra1.6 μm	4+1	超差 0.01 mm 扣 2 分			
13		内径	$42\phi_{0}^{+0.03}$ mm/Ra1.6 μm	4+1	超差 0.01 mm 扣 2 分			
14			$40\phi_{0}^{+0.03}$ mm/Ra1.6 μm	4+1	超差 0.01 mm 扣 2 分			
15			$\phi36_{0}^{+0.03}$ mm/Ra1.6 μm	4+1	超差 0.01 mm 扣 2 分			
16			$\phi27_{0}^{+0.03}$ mm/Ra1.6 μm	4+1	超差 0.01 mm 扣 2 分			
17		长度	2 mm	2	超差 0.2 mm 全扣			
18			7 mm	2	超差 0.2 mm 全扣			
19			13 mm	2	超差 0.2 mm 全扣			
20			15 mm	2	超差 0.2 mm 全扣			
21			24 mm	2	超差 0.2 mm 全扣			
22			48±0.1 mm	2	超差 0.2 mm 全扣			
23		圆弧	$R3$ mm/Ra1.6 μm	2+1	目测			
24	配合间隙		5±0.1 mm	4	超差全扣			
25	配合长度		103±0.1 mm	4	超差全扣			
			78±0.1 mm	4	超差全扣			
26	圆锥配合		接触面 60% 以上	4	目测			
27	工艺、程序		遵守工艺与程序的有关规定		违反规定扣总分 1~5 分			
28	规范操作		遵守数控车床规范操作的有关规定		违反规定扣总分 1~5 分			
29	安全文明生产		遵守安全文明生产的有关规定		违反规定扣总分 1~5 分			
学生任务实施过程的小结及反馈：								
教师点评：								

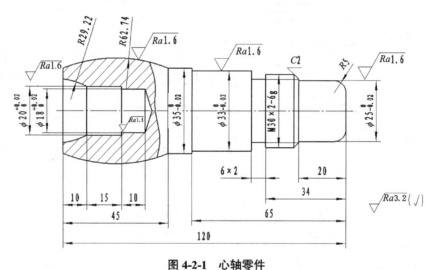

::: center
任务二　　模具心轴零件加工
:::

一、图样与技术要求

　　根据图 4-2-1 所示的心轴，制定加工方案，在数控车床上完成加工。毛坯：$\phi50$ mm×125 mm 棒料。材料：45 钢。

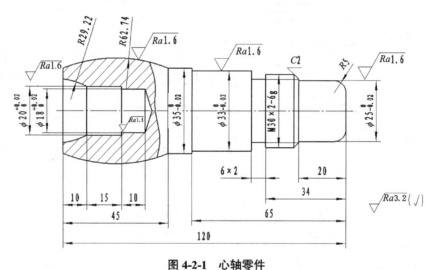

::: center
图 4-2-1　心轴零件
:::

二、图纸分析

1. 图纸分析

　　如图 4-2-1 所示心轴零件图，该零件的轮廓描述清楚，尺寸标注完整。从轮廓上看该零件在外部轮廓上有台阶、槽、螺纹、圆弧，内部轮廓则有内接台与内圆弧，

　　1）心轴的概念

　　心轴是用来支撑转动的零件，只承受弯矩而不传递扭矩。图 4-2-2 所示为心轴。

图 4-2-2　心轴

2）心轴分类

根据轴工作时是否转动，心轴又可分为转动心轴和固定心轴。

（1）转动心轴：工作时轴承受弯矩，且轴转动，如铁路车辆的轴等。

（2）固定心轴：工作时轴承受弯矩，且轴固定，如支撑滑轮的轴等。

2. 工艺安排

1）确定工件的装夹方案

图 4-2-1 所示心轴零件左端内轮廓长为 35 mm，外轮廓为 $R62.74$ mm 的圆弧表面、长为 45 mm，右端为台阶轴，零件总长 120 mm。鉴于零件的总体形状，加工右部的 $\phi25$ mm、$\phi30$ mm、$\phi33$ mm、$\phi35$ mm 台阶时可采用一夹一顶的方式装夹（在工件左端可做 $\phi46$ mm×10 mm 工艺台）。对于右端的装夹可采用三爪自定心卡盘进行装夹加工。

2）确定加工路线

在此次加工任务中，尽可能做到由内到外、由粗到精、由近到远以及在一次装夹中加工出较多的表面，在加工内、外轮廓时，可加工内孔各表面。具体加工顺序如下。

（1）平右端面，打中心孔并光整毛坯外圆表面长度 75 mm。

（2）掉头找正，三爪卡盘夹持已光整外圆，车 $\phi47$ mm 外圆，长度 10 mm 工艺台，做总长 120.5 mm。

（3）采用一夹一顶的装夹方式，粗车加工 $\phi35_{-0.02}^{0}$ mm 外圆、$\phi33_{-0.02}^{0}$ mm 外圆、$\phi30_{-0.02}^{0}$ mm 外圆、$\phi25_{-0.02}^{0}$ mm 外圆及 $R5$ mm 圆弧，径向留量 0.5 mm，轴向留量 0.1 mm，轴向长度 75 mm 。

（4）切槽刀加工 6 mm×2 mm 螺纹退刀槽。

（5）精车 $\phi35_{-0.02}^{0}$ mm 外圆、$\phi33_{-0.02}^{0}$ mm 外圆、$\phi30_{-0.02}^{0}$ mm 外圆、$\phi25_{-0.02}^{0}$ mm 外圆及 $R5$ mm 圆弧至图纸要求。

（6）螺纹刀车削 M30×2-6g 外螺纹至精度要求。

（7）掉头采用三爪卡盘装夹，垫铜皮夹持 $\phi33_{-0.02}^{0}$ mm 外圆找正夹紧，平端面做总长 120 mm。

（8）打 B2 中心孔。

（9）$\phi16$ mm 钻头钻孔，内孔长 35 mm。

（10）粗镗内部轮廓其中包括 $R29.22$ mm 内圆弧、$\phi20$ mm 内孔、$\phi18$ mm 内孔。

（11）精镗以上各部尺寸至图纸要求。

（12）粗、精车 *R*62.74 mm 外圆弧至图纸要求。

3）填写加工刀具卡和工艺卡

加工刀具卡和工艺卡，见表4-2-1。

表4-2-1 加工刀具卡和工艺卡

零件图号		4-2-1	数控车床加工工艺卡	机床型号	CKA6150
零件名称		模具心轴零件		机床编号	
刀具表				量具表	
刀具号	刀具补偿号	刀具名称	刀具参数	量具名称	规格 /mm
T01	01	90°外圆粗、精车刀	C 型刀片	游标卡尺 千分尺	0~150/0.02 25~50/0.01
T02	02	93°外圆精、精车刀	D 型刀片	游标卡尺 千分尺	0~150/0.02 25~50/0.01
T03	03	切槽刀	刃宽 4.0 mm	游标卡尺	0~150/0.02
T04	04	外螺纹刀	刀尖角 60°	环规	M30×2
T05	05	91°镗孔车刀	D 型刀片	内径百分表	18~35/0.01
		中心钻	B2		
		钻头 φ16 mm		游标卡尺	0~150/0.02

工序		工艺内容	切削用量			加工性质
			S/（r/min）	*F*/（mm/r）	*a*p/mm	
数控车	1	三爪卡盘装夹，车外圆、端面确定基准	600	—	1	手动
	2	粗加工 φ35 mm、φ33 mm、φ30 mm、φ25 mm外圆及 *R*5 mm 圆弧	500~700	0.2	2	自动
	3	切 6 mm×2 mm 的退刀槽	400~600	0.05	—	自动
	4	精加工 φ35 mm、φ33 mm、φ30 mm、φ25 mm外圆及 *R*5 mm 圆弧	1 000~1 200	0.1	0.25	自动
	5	加工 M30×2 外螺纹	600	2.0	—	自动
数控车	1	掉头装夹找正保证总长 120 mm	500	0.1		手动
	2	手动钻中心孔	800	—		手动
	3	钻φ16 mm孔	300	—		手动
	3	粗镗内部轮廓其中包括（*R*29.22 mm 内圆弧、φ20 mm、φ18 mm、内孔）	500~700	0.15	1.5	自动
	4	精镗内部轮廓其中包括（*R*29.22 mm 内圆弧、φ20 mm内孔、φ18 mm、内孔）	700~1000	0.08	0.25	自动
	5	粗车 *R*62.74 mm 外圆弧。	600~700	0.2	1	自动
	6	精车 *R*62.74 mm 外圆弧至图纸要求。	1 000~1 200	0.1	0.25	自动

三、程序编制与零件加工

根据图 4-2-1 所示零件，分析工件加工要求，确定加工装夹方案、加工路线及采用的刀具和切削用量，根据工艺过程将工序内容划分为三个部分，对应编制三个程序并完成加工。表 4-2-2 为加工零件右端外形轮廓程序，表 4-2-3 为加工零件左端内部轮廓程序，表 4-2-4 为加工零件左端外形轮廓程序。

表 4-2-2 加工零件右端外形轮廓程序

程序内容	程序说明
O4201；	程序号；
N1；	第 1 程序段；
G99 M03 S600 T0101；	选 1 号刀，主轴正转，转速 600 r/min；
G00 X200.0 Z10.0；	快速运动到安全点；
G00 X50.0 Z2.0；	快速运动到循环点；
M08；	冷却液开；
G71 U2.0 R0.5；	粗加工循环；
G71 P10 Q20 U0.5 W0.05 F0.2；	
N10 G00 X15.0；	循环加工起始程序段；
G01 Z0.0 F0.1；	
G02 X25. Z–5.0 R5.0；	
G01 Z–20.0；	
G01 X26.0	
G01 X29.9 W–2.0；	
G01 Z–40.0；	
G01 X33.0；	
G01 Z–65.0；	
G01 X35.0，	
G01 W–10.0；	
N20 G01 X50.0；	循环加工终点程序段；
G00 X200.0；	退刀；
G00 Z10.0；	
M09；	冷却液关；
M05；	主轴停；
M00；	程序暂停；
N2；	第 2 程序段；
G99 M03 S500 T0303；	选 3 号刀（切刀），主轴正转，转速 500 r/min；
G00 X35.0 Z–40.0；	快速运动到起刀点；
M08；	冷却液开；
G01 X26.0 F0.05；	切槽；
G00 X35.0；	退刀；
G00 W2.；	轴向移动 2 mm；
G01 X26.0 F0.05；	切槽；
G00 X200.0；	径向退刀；
G00 Z10.；	轴向退刀；
M09；	冷却液关；
M05；	主轴停；
M00；	程序暂停；

续表

程序内容	程序说明
N3；	第3程序段；
G99 M03 S1200 T0101；	选1号刀（90°外圆正偏刀），主轴正转，转速1 200 r/min；
G00 G42 X50.0 Z2.0；	调刀置循环起点，添加刀尖圆弧右补偿；
M08；	冷却液开；
G70 P10 Q20；	精加工循环；
G00 G40 X200.0 Z10.0；	快速运动到安全点，取消刀尖圆弧半径补偿；
M09；	冷却液关；
M05；	主轴停；
M00；	程序暂停；
N4；	第4程序段；
G99 M03 S600 T0404；	选4号刀（外螺纹刀），主轴正转，转速600 r/min；
G00 X35.0 Z−15.0；	快速运动到循环起刀点；
M08；	冷却液开；
G92 X29.2 Z−37.0 F2.0；	螺纹切削循环；
X28.6；	
X28.1；	
X27.7；	
X27.5；	
X27.4；	
G00 X200.0 Z10.0；	退刀；
M09；	冷却液关；
M05；	主轴停；
M30；	程序结束返回第一程序段

表4-2-3　加工零件左端内部轮廓程序

程序内容	程序说明
O4202；	程序号；
N1；	第1程序段；
G99 M03 S600 T0505；	选5号刀，主轴正转，转速600 r/min；
G00 X100.0 Z100.0；	快速运动到安全点；
G00 X16.0 Z2.0；	快速运动到循环点；
M08；	冷却液开；
G71 U1.5 R0.5；	粗加工循环；
G71 P10 Q20 U−0.5 W0.05 F0.15；	
N10 G00 X26.06；	加工起始程序段；
G01 Z0.0 F0.1；	
G03 X20.0 Z−10.0 R29.22；	
G01 W−15.0；	
G01 X18.0；	
G01 W−10.0；	
N20 G01 X16.0；	循环加工终点程序段；
G00 Z300.0；	退刀；
G00 X100.0；	
M09；	冷却液关；
M05；	主轴停；
M00；	程序暂停；

程序内容	程序说明
N2;	第 2 程序段;
G99 M03 S1200 T0505;	选 5 号刀,主轴正转,转速 1 200 r/min;
G00 X100.0 Z100.0;	快速运动到安全点;
G00 G41 X16.0 Z2.0;	快速运动到循环点加刀尖圆弧半径左补偿;
M08;	冷却液开;
G70 P10 Q20;	精加工循环;
G00 Z300.0;	退刀;
G00 G40 X200.0;	取消刀具补偿;
M09;	冷却液关;
M05;	主轴停;
M30;	程序结束返回第一程序段

表 4-2-4　加工零件左端外形轮廓程序

程序内容	程序说明
O4203;	程序号;
N1;	第 1 程序段;
G99 M03 S600 T0202;	选 2 号刀,主轴正转,转速 600 r/min;
G00 X100.0 Z100.0;	快速运动到安全点;
G00 X50.0 Z2.0;	快速运动到循环点;
M08;	冷却液开;
G71 U1.0 R0.5;	粗加工循环;
G71 P10 Q20 U0.5 W0.05 F0.2;	
N10 G00 X35.0;	循环加工起始程序段;
G01 Z0. F0.1;	
G03 X35.0 Z–45.0 R52.74;	
N20 G01 X50.0;	循环加工终点程序段;
G00 X100.0 Z100.0;	快速运动到安全冷却液关;
M05;	主轴停;
M09;	冷却液关;
M00;	程序暂停;
N2;	第 2 程序段;
G99 M03 S1200 T0202;	选 2 号刀,主轴正转,转速 1 200 r/min;
G00 G42 X50.0 Z2.0;	快速运动到循环点,刀具右补偿;
M08;	冷却液开;
G70 P10 Q20;	精加工循环;
G00 Z300.0;	快速运动到安全点;
G00 G40 X200.0;	取消刀具补偿;
M09;	冷却液关;
M05;	主轴停;
M30;	程序结束返回第一程序段

四、任务评价

按表 4-2-5 进行任务评价。

表 4-2-5 心轴零件评分标准

	考核项目	考核要求	配分	评分标准	检测结果	得分	备注
1		$\phi 42_{-0.02}^{0}$ mm	8	超差 0.01 mm 扣 4 分			
2	外径	$\phi 33_{-0.02}^{0}$ mm	8	超差 0.01 mm 扣 4 分			
3		$\phi 35_{-0.02}^{0}$ mm	8	超差 0.01 mm 扣 4 分			
4	内径	$\phi 20_{0}^{+0.02}$ mm	8	超差 0.01 mm 扣 4 分			
5		$\phi 18_{0}^{+0.02}$ mm	8	超差 0.01 mm 扣 4 分			
6		$R5$ mm	8	R 规检测			
7	圆弧	$R62.74$ mm	8	R 规检测			
8		$R29.22$ mm	8	R 规检测			
9	螺纹	M30×2–6g	14	螺纹通、止规检测			
10	长度	8 处	8	超差全扣			
11	表面结构	8 处 $Ra1.6$ μm	8	Ra 值大 1 级无分			
12	退刀槽	6 mm×2 mm	6	超差全扣			
13	工艺、程序	遵守工艺与程序的有关规定		违反规定扣总分 1~5 分			
14	规范操作	遵守数控车床规范操作的有关规定		违反规定扣总分 1~5 分			
15	安全文明生产	遵守安全文明生产的有关规定		违反规定扣总分 1~5 分			

学生任务实施过程的小结及反馈：

教师点评：

Practice for CNC Lathe Operation and Programming

WU YUNFEI

天津大学出版社

TIANJIN UNIVERSITY PRESS

Preface

The *Practice for CNC Lathe Operation and Programming* was prepared to coordinate with the theoretic and practical training and teaching of "Luban Workshop" in Madagascar, carry out exchanges and cooperation, improve the international influence of China's vocational education, innovate the international cooperation mode of vocational schools, and export excellent vocational education resources of China on the basis of the implementation of the cooperation on vocational education between Tianjin and the countries along the Belt and Road.

This book is written in both Chinese and English for Luban Workshop project in Madagascar. With numerical control machining equipment of Luban Workshop as the carrier, it reflects the position needs on the basis of actual situation of enterprises and in compliance with the National Occupational Standards and the concept of "taking occupational standards as basis, enterprises needs as orientation, and occupational ability as the core", highlighting the new knowledge, new technology, new process, and new method, and centering on the training of occupational ability. There are four items and 11 tasks in total, trying to explain the profound things in a simple way and facilitate teaching.

The content of this book goes from simple to deep, combine theory with practice, and arranges knowledge points and skill points together in task-driven form with appropriate cases. Based on the principles of scientificity, practicability and universality, this book is in line with the status quo of the mechanical curriculum system of vocational education. This book focuses on the turning operation and programming method of FANUC 0itd CNC system and gives a complete list of machining procedures and relevant descriptions in combination with examples. Precautions, skills and other aspects for programming are explained with small columns such as "Note" and "Attention". This book is available for engineers and operators using the FANUC 0i system, as well as for CNC teachers and students of various vocational colleges and contestants of the CNC technology competition.

This book was edited by Wu Yunfei of Tianjin Technician Institute of Mechanical & Electrical Technology and reviewed by Xu Guosheng of Tianjin University of Technology and Education. Xu Chunnian, Yang Fuling and Zang Chengyang of the Tianjin Technician Institute of Mechanical & Electrical Technology also participated in the preparation. Item I was compiled by Xu Chunnian, Item II by Yang Fuling, Item III by Zang Chengyang, and

Item IV by Wu Yunfei. We would like to express our gratitude for the great help of CBS in the preparation of this book！ At the same time，we have referred to a number of publicly published literature at home and abroad，and hereby thank the authors！

Limited by the experience and level of the editor，defects are unavoidable in this book，please do not hesitate to raise your valuable opinion.

Editors

Contents

Item 4   Programming and Machining of Complex Parts ·············· 107

Basis of CNC Lathe Operation

[Item description]

CNC machine tool is cold working equipment that takes digital operation as the core, executes program code, and has a computer to coordinate various components so as to process parts. It is one of the core technologies of intelligent manufacturing. As high-speed, high-precision and high-reliability machinery processing equipment, it is mainly used in IT, automotive, light industry, medical, aviation and other industries. Learning this part should integrate theory with practice. While introducing basic CNC knowledge, use multimedia equipment to enable students to have a more intuitive understanding of CNC machine tools and CNC systems, and perform appropriate operation demonstrations to deepen their impression.

[Learning objectives]

（1）Understand the safe operation rules, the maintenance and repair, and the generation and composition of CNC lathe.

（2）Familiar with the relevant definitions and classifications of CNC machine tools.

（3）Master the main technical specifications and parameters of CK6150e CNC horizontal lathe and the manual operation of CNC lathe.

（4）Familiar with FANUC 0i CNC lathe panel.

（5）Master the definition of lathe coordinate system and basic points.

（6）Master the basic meaning of various codes.

Task I Introduction education-knowledge of CNC lathe

I. Task description

Complex shaft parts are often encountered in production and processing. In order to ensure processing accuracy and improve production efficiency, CNC lathes are generally selected for processing. Therefore, it is necessary for students to learn about the equipment they use and preliminarily understand the production tasks by reading materials and observing operations before they operate the CNC lathe independently.

II. Task analysis

By reading machine tool instructions, textbooks and downloading relevant materials on the Internet, students shall have a good command of the safe operation rules, the maintenance and repair of the CNC lathe to prepare for the formal operation of the machine tool.

III. Knowledge and skills

CNC equipment is advanced machining equipment with a high degree of automation and complex structure and is the key equipment of enterprises. To give full play to the high efficiency of CNC equipment, correct operation and careful maintenance are necessary to ensure the utilization rate of the equipment. Correct operation and use can prevent abnormal wear of the machine tool and avoid sudden failure; Daily maintenance and repair can keep the equipment in good technical condition, delay the deterioration process, timely detect and eliminate the hidden dangers of failure, thus ensuring safe operation.

1. Safe operation rules of the CNC lathe

(1) Carefully read the instruction manuals and relevant operation handbooks before operating the machine tool. It is strictly prohibited to modify the maker setting parameters of CNC and PMC without authorization. Do not move or damage the warning signs installed on the machine tool.

(2) Before startup, the CNC lathe shall be comprehensively and carefully inspected, including operation panel, guide rail surface, claw, tailstock, tool post, tool, etc., and then it can be operated after confirmation.

(3) After the CNC lathe is powered on, check whether the switches, buttons and keys are normal and flexible and whether the machine tool has any abnormalities.

(4) Before the machine tool is started, it is necessary to idle the machine tool at medium and low speed to preheat it for more than 5 minutes, and carefully check whether the lubrication system works normally. If the machine tool has not been started for a long time, each part first must be manually supplied oil for lubrication.

(5) After the program is entered, carefully check whether the code, address, value,

positive and negative numbers, decimal points and syntax are correct.

（6）Correctly measure and calculate the workpiece coordinate system and check the results obtained.

（7）Enter the workpiece coordinate system and carefully check the coordinates, coordinate values, positive and negative numbers and decimal points.

（8）Before installing the workpiece, idle the program once to see whether the program can run smoothly, whether the tool and fixture are installed reasonably and whether there is overrun. The tool used must match the specifications of the machine tool.

（9）Check the status of chuck clamping. Before the machine tool is started, the machine tool protective door must be closed. It is forbidden to touch the tool tip and iron filings by hand. The iron filings must be cleaned with an iron hook or brush. Do not touch the rotating spindle, workpiece and other moving parts by hand or by any other means.

（10）The fast rate switch must be adjusted to the lower gear during the trial cut. When the workpiece extends 100mm beyond the rear end of the lathe spindle, protective objects must be set at the extended position.

（11）During the trial cutting feed, when the tool runs to 30-50mm of the workpiece, it is necessary to verify whether the residual values of Z-axis and X-axis coordinates are consistent with the machining program under the feed hold.

（12）During trial cutting and machining, after grinding the tool and changing the tool, it's necessary to re-measure the tool position and modify the tool offset value and offset number.

（13）After the program is modified, the modified part shall be carefully checked.

（14）In case of abnormal conditions such as workpiece runout, shaking, abnormal sound and fixture looseness during operation, stop the machine for treatment.

（15）After emergency shutdown, the machine tool "zero return" operation shall be carried out again before the program can be run again.

（16）Only one person is allowed to operate the machine tool, and two or more persons are not allowed to operate equipment at the same time. The operator shall stop when leaving the machine tool, changing the speed, replacing the tool, measuring the size and adjusting the workpiece.

（17）After the operation, the machine tool shall be "returned to zero" and carefully cleaned. Compressed air is not allowed to clean the machine tool, electrical cabinet and NC unit.

2. Maintenance and repair of CNC lathe

The CNC lathe has the characteristics of technology-intensive and knowledge-intensive that integrates machinery, electricity and liquid. It is advanced processing equipment with high degree of automation, complex structure and high cost. In order to give full play to its benefits and reduce the occurrence of faults, daily maintenance must be done well. Therefore, the maintenance personnel of CNC lathe is required to have not only the knowledge of machinery, processing technology, hydraulic pressure and pneumatic, but also the knowledge of electronic computer, automatic control, driving and measurement technology, so as to fully understand and master the CNC lathe and do a good job in maintenance in time. The maintenance work includes the following:

（1）Select a suitable service environment. The service environment of the CNC lathe (such as temperature, humidity, vibration, power voltage, frequency and interference) will affect the normal operation of the machine tool, so the installation conditions and requirements specified in the machine tool instructions shall be strictly met when installing the machine tool. When economic conditions permit, CNC lathes and ordinary machining equipment shall be installed separately to facilitate repair and maintenance.

（2）The CNC lathe shall be equipped with specialized personnel for CNC system programming, operation and maintenance. These personnel shall be familiar with the machinery, CNC system, strong current equipment, hydraulic pressure, air pressure and other parts of the machine tool used, understand the requirements of equipment use environment and machining conditions, and be able to correctly use the CNC Lathe according to the instruction manuals of the machine tool and system.

（3）Servo motor maintenance The servo motor of the CNC lathe shall be maintained once every 10-12 months, and the machine tool with frequent acceleration or deceleration changes shall be maintained once every 2 months. The main contents of maintenance include: blow off the dust on the brush with dry compressed air, check the wear of the brush, if replacement is required, select the brush of the same specification and model, and run it without load for a certain period of time after replacement to match the surface of the commutator; Check and clean the armature commutator to prevent short circuit. If the speed measuring motor and pulse encoder are installed, check and clean them.

（4）Timely cleaning For example, air filter cleaning, electrical cabinet cleaning and printed circuit board cleaning.

（5）The inspection of machine tool cables mainly checks whether there are poor contact, disconnection and short circuit at the moving joints and corners of the cables.

（6）The parameter memory of some CNC systems adopts CMOS elements, and its storage content is maintained by battery power supply in case of power failure. Generally, the battery should be replaced once a year, and it must be used when the CNC system is powered on, otherwise, the stored parameters will be lost, resulting in the failure of the CNC system.

（7）Maintenance of CNC lathe not used for a long time When the CNC lathe is left unused，the CNC system should be powered on frequently to make it run idle with the machine tool locked. During the plum rain season with high air humidity，the CNC system should be powered on every day so that the heat of the electrical components themselves can remove the moisture in the electrical cabinet to ensure stable and reliable performance of the electrical components.

See Table 1-1-1 for the maintenance schedule of one CNC lathe.

Table 1-1-1　Maintenance of CNC Lathe

S/N	Inspection cycle	Inspection part	Inspection requirements
1	Daily	Guide rail lubricating oil tank	Check the oil volume, add lubricating oil in time, and check whether the lubricating oil pump starts and stops regularly
2	Daily	Spindle lubrication thermostatic oil tank	Check whether the operation is normal, the oil volume is sufficient and the temperature range is appropriate
3	Daily	Hydraulic system of machine tool	Check whether the oil tank pump has abnormal noise, whether the working oil level is appropriate, whether the indication of the pressure gauge is normal, and whether the pipeline and joints have leakage
4	Daily	Pressure of compressed air source	Check whether the pressure of the pneumatic control system is within the normal range
5	Daily	Guide rail surface of X and Z axes	Remove cuttings and dirt, check whether the guide rail surface is scratched or damaged and whether the lubricating oil is sufficient
6	Daily	Protective devices	Check whether the machine tool protective cover is complete and effective
7	Daily	Heat dissipation and ventilation devices of the electrical cabinet	Check whether the cooling fan in each electrical cabinet works normally, whether the air duct filter is blocked, and clean the filter in time
8	Every week	Filter screens of each electrical cabinet	Clean the adhesive dust
9	Irregular	Coolant tank	Check the level at any time, add coolant in time, and replace it if it is too dirty
10	Irregular	Chip remover	Clean the cuttings frequently and check for jamming
11	Half a year	Check the spindle drive belt	Adjust the belt tightness according to the instructions

Continue Table

S/N	Inspection cycle	Inspection part	Inspection requirements
12	Half a year	Strips and pressing rollers on guide rails of each axle	Adjust the tightness according to the instructions
13	One year	Check and replace the motor carbon brush	Check the commutator surface, remove burrs, blow off carbon powder, and replace the worn carbon brush in time
14	One year	Hydraulic oil line	Clean the overflow valve, pressure reducing valve, oil filter, oil tank, filter hydraulic oil or replace them
15	One Year	Spindle lubrication thermostatic oil tank	Clean the filter, oil tank and replace the lubricating oil
16	One year	Cooling oil pump filter	Clean the cooling oil pool and replace the filter
17	One year	Ball screw	Clean the old grease on the screw and apply new grease

IV. Task implementation

（1）Students understand and familiarize themselves with the safe operation rules and the maintenance and repair of the CNC machine tools through reading materials.

（2）Enter the training workshop to consolidate the known knowledge through observation and operation.

V. Task assessment

Assessment of CNC lathe safe operation and machine tool maintenance.

（1）Assessment form: oral examination.

（2）Assessment contents:

① Safe Operation rules of CNC Lathe;

② Student practice code;

③ Briefly describe the routine inspection contents of CNC lathe.

（3）Assessment requirements: Simply and concisely describe the above problems and highlight the key points.

Task II Overview and manual operation of CNC lathe

I. Task description

CNC lathe is one of the widely used CNC machine tools at present. It is mainly used for cutting the internal and external cylindrical surfaces of shaft parts and disk parts, internal and external tapered surfaces with arbitrary taper angle, complex rotating internal and external curved surfaces, cylinders and tapered threads, and also can be used for cutting grooves, drilling, reaming and boring. CK6150e is a new generation of economical CNC lathe which adopts the FANUC 0i TD system, as shown in Fig. 1-2-1.

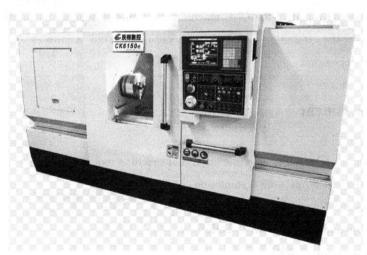

Fig. 1-2-1 CK6150e CNC Lathe

The machine tool is arranged horizontally and the CNC system controls the movement of horizontal (X) and longitudinal (Z) coordinates. It mainly undertakes semi-finishing and finishing of various shaft and disc parts. Internal and external cylindrical surfaces, tapered surfaces, threads, holes and various curve cursors can be machined with it. The spindle box adopts a variable frequency motor to realize manual three gear stepless speed regulation, and the tool post quantity is 4, which is suitable for teaching and enterprise manufacturing.

II. Knowledge and skills

1. Introduction to CNC Lathe

The numerical control machine tool, also known as the CNC (Computer Numerical Control) machine tool, is a machine tool installed with a program control system that can logically process the program specified by the use number or other symbol coding instructions. Digital control is an automatic control technology developed in modern times. It uses digital information to control an object which can be displacement, speed, temperature, pressure, flow, color and so on.

2. Composition and classification of CNC lathe

1) Composition of CNC lathe

There are many types of CNC lathes, which are generally composed of the lathe body, CNC device and servo system. Fig. 1-2-2 is a schematic diagram of the basic composition of the CNC lathe.

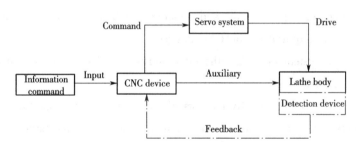

Fig. 1-2-2 Schematic Diagram of Basic Composition of CNC Lathe

(1) Lathe body. Lathe body is the actual mechanical part to realize the machining process which mainly includes: main moving parts (such as chuck, spindle, etc.), feed moving parts (workbench, tool post, etc.), supporting parts (bed, column, etc.) and auxiliary devices such as cooling, lubrication and rotating parts and clamping and tool changing manipulators.

The main body of the CNC lathe is specially designed, and the performance of each part is superior to that of ordinary lathe. For example, the structure is rigid that can meet the needs of high-speed turning; High precision and good reliability can adapt to precision machining and long-time continuous work.

(2) CNC device and servo system. The main difference between the CNC lathes and ordinary lathes is whether they have two parts: CNC device and servo system. If we compare the detection device of the CNC lathe to human eyes, then the CNC device is equivalent to

the human brain and the servo system is equivalent to human hands. Therefore, it can be seen that how important the two parts are in the CNC lathe.

① CNC device.The core of the CNC device is the computer and the software running on it. It plays a "command" role in the CNC lathe, as shown in Fig. 1-2-3. The CNC device receives information sent by the machining program and sends an execution command to the drive mechanism after processing and deployment. During execution, mechanisms like drive and detection feedback relevant information to the CNC device at the same time, so that a new execution command can be issued after processing.

Fig. 1-2-3　CNC Device

② Servo system.The servo system accurately executes the commands issued by the CNC device through the drive circuit and actuator (such as the servo motor) to complete various displacements required by the CNC device.

The feed drive system of CNC lathe often works with feed servo system, so it is also called feed servo system.

The feed servo system generally consists of five parts: position control, speed control, servo motor, detection component and mechanical transmission mechanism. However, the feed servo system, as usual, refers to only three parts: speed control, servo motor and detection component. In addition, the speed control part is called a servo unit or drive.

2) Classification of CNC lathe

There are many types of CNC lathes, and the commonly used lathes are classified as follows.

(1) Classification by spindle distribution form of the CNC lathe.

① Vertical CNC lathe The spindle of the vertical CNC machine tool is perpendicular to the horizontal plane, and there is a worktable with a large diameter for clamping the workpiece. It is mainly used to process large and complex parts with large radial dimensions and small axial dimensions, as shown in Fig. 1-2-4.

② Horizontal CNC lathe The spindle of the horizontal CNC machine tool is parallel to the horizontal plane, which can be further divided into horizontal guide rail horizontal CNC lathe and inclined guide rail horizontal CNC lathe. Fig. 1-2-5 shows the horizontal CNC lathe with an inclined guide rail.

Fig. 1-2-4　Vertical CNC Lathe

Figure 1-2-5　Horizontal CNC Lathe

（2）Classification by number of tool posts.

① single tool post CNC lathe, as shown in Fig. 1-2-6.

② double tool post CNC lathe, as shown in Fig. 1-2-7.

Fig. 1-2-6　Single Tool
Post CNC lathe

Fig. 1-2-7　Double Tool Post
Horizontal CNC Lathe

（3）Classification by CNC System Function.

① Economical CNC lathe This type of lathe is often based on the CNC transformation of ordinary lathes. Generally, they have front tool posts. They are mainly used for the processing of parts with low machining accuracy and certain complex shapes, as shown in Fig. 1-2-1.

② Full-feathered CNC lathe As shown in Fig. 1-2-5, this kind of lathe has advanced overall structure, complete control functions, high degree of machining automation, complete auxiliary functions, good stability and reliability. It is suitable for machining parts with high precision and complex shapes.

③ CNC turning center As shown in Fig. 1-2-8, the turning milling center is a machine tool with a full-feathered CNC lathe as the main body and equipped with tool magazines, tool changer, indexing device, milling power head and manipulator, etc., which can realize multi-process composite machining. After the parts are clamped at one time, various processes such as turning, milling, drill, expand, reaming, tapping screw thread and so on can be completed.

Fig. 1-2-8　CNC lathe milling center

III. Task implementation

1. Main technical specifications and parameters of CK6150e CNC horizontal lathe are shown in Table 1-2-1

Table 1-2-1　The Main Technical Parameters of CK6150e Horizontal CNC Lathe

	Item	Parameter Value
Technical specifications	Maximum swing diameter over bed/mm	500
	Maximum workpiece swing diameter over sliding plate/mm	280
	Maximum machining length/mm	1 000
Main drive (Variable frequency stepless speed regulation)	Spindle speed step	Stepless
	Spindle speed range/ (r/min)	200-2 200
	Spindle end structure	C8
	Diameter of spindle hole	$\phi 82$
	Front-end taper of spindle hole	90 metric 1 : 20
	Main motor model	YVP132M-4-7.5 kW
Feed system	Maximum travel of tool post/mm	Direction OX: 300; direction OZ: 1 050
	Rolled ball screw diameter x pitch/mm	Direction OX: $\phi 25 \times 5$ direction OZ: $\phi 40 \times 6$
	Fast moving feed/ (mm/min)	Direction OX: 8 000; direction OZ: 10 000
	Positioning accuracy/mm	Direction OX: 0.015; direction OZ: 0.020
	Repeated positioning accuracy/mm	Direction OX: 0.010; direction OZ: 0.015
	Workpiece machining accuracy	IT6-IT7
	Surface roughness of workpiece	$Ra1.6$ μm

Continue Table

Item		Parameter Value
Tailstock device	Diameter of tailstock sleeve/mm	75
	Travel of tailstock sleeve/mm	170
	Taper of taper hole of tailstock sleeve	Morse 5°
Tool post device	Standard configuration	Vertical 4 position electrical tool post
	Special selection configuration	Six/eight position electric tool post
	Repeated positioning accuracy/mm	0.005
	Section of tool rod/mm	25×25
	Select configuration	FANUC, Siemens, Central China, KND
Overall dimension and weight of machine tool	Appearance dimension of machine tool (length × width × height) /mm	2 800×1 600×1 870
	Net weight of machine tool/kg	2 800

2.Manually operated machine tool

When the machine tool automatically machines the workpiece according to the machining procedure, the machine tool operation is basically automatic, while in other cases, the machine tool is operated manually.

(1) Since the incremental measuring system is adopted for manual return to the machine tool reference point, once the machine tool is powered off, the CNC system loses the memory of the reference point coordinates. When the CNC system is powered on again, the operator must first make the machine tool return to the reference point. In addition, when the machine tool encounters an emergency stop signal or an over-travel alarm signal during operation, the operation of returning to the reference point must also be carried out when the machine tool resumes operation after the fault is eliminated. The specific operation steps are as follows: set the "MODE" switch to ZERO RETURN mode. Remind the operator that when the distance between the stopper on the sliding plate and the reference point switch is less than 30 mm, first use the "JOG" button to move the sliding plate towards the negative direction of the reference point until the distance is more than 30 mm, stop moving, and then return to the reference point. Press the "JOG" button on the X-axis and Z-axis respectively to move the sliding plate forward to the reference point along the X-axis or Z-axis. During this process, the operator shall press the "JOG" button until the reference point returns and the indicator light is on, and then release the button. When the sliding plate moves near the two-axis reference point, it automatically slows down.

(2) Manual feed of the sliding plate is required when the machine tool is manually adjusted or when the tool is required to move quickly to approach or leave the workpiece.

There are two manual operations for slide feed: one is to quickly move the sliding plate with the "JOG" button and the other is to move the sliding plate with the handwheel.

(3) When moving the tool quickly or operating manually, it is required that the tool can move quickly to approach or leave the workpiece. The operation is as follows: put the "MODE" switch in the "RAPID" mode; Use the "APIDOVERRIDE" switch to select the speed at which the sliding plate moves fast; Press the "JOG" button to quickly move the tool post to the predetermined position.

(4) When manually adjusting the tool with the handwheel feed, use the handwheel to determine the correct position of the tool tip, or during trial cutting, use the handwheel to subtly tune the feed speed while observing the cutting condition. The operating steps are as follows: turn the "MODE" switch to the "HANDLE" position (3 positions can be selected), and select the movement amount of the sliding plate for each 1 grid of rotation of the handwheel. Turn the "MODE" switch to ×1 and rotate the handwheel 1 grid to move the sliding plate 0.001 mm; If it points to ×10, rotate the handwheel 1 grid to move the sliding plate 0.01 mm; If it points to ×100, rotate the handwheel 1 grid to move the sliding plate 0.1 mm; Pull the X-axis and Z-axis selection switches on the left side of the handwheel to the coordinate axis to be moved by the sliding plate; Turn the manual pulse generator to make the tool post move in the specified direction and speed.

(5) The operation of the spindle mainly includes start, stop and inching of the spindle.

① The start and stop of the spindle are used to adjust the tool or debug the machine tool. The specific operating procedure is to place the "MODE" switch at any position in the manual mode (MANU). Use the "FWD-RVS" switch in the spindle function button to determine the rotation direction of the spindle. In the "FWD" position, the spindle rotates forward; In the "RVS" position, and the spindle rotates reversely. Rotate the spindle "SPEED" knob to the low-speed zone to prevent sudden acceleration of the spindle. Press the "START" button to rotate the spindle. During the spindle rotation, the spindle speed can be controlled by the "SPEED" knob, and the actual spindle speed is displayed on the CRT display. Press the spindle STOP button can stop the spindle rotation.

② The inching of the spindle is used to rotate the spindle to a position convenient for handling the jaws or for checking the clamping of the workpiece. It is operated by placing the "MODE" switch at any position in the automatic mode (AUTO). Point the spindle "FWD-RVS" switch to the desired rotation direction. Press the "start" button to rotate the spindle; When the button is lifted, the spindle stops rotating.

(6) The rotation of tool post, loading and unloading of tools, measuring the position of cutting tools and trial cutting of workpiece shall be programmed and executed in MDI state. The operating procedure is to put the "MODE" switch in the "MDI" mode; press the

function key "PRGRM" to enter T10/T20/T30/T40 and then press the START key.

（7）Operation of the manual tailstock includes movement of the tailstock body and movement of the tailstock sleeve.

① Movement of the tailstock body It is mainly used to adjust the position of the tailstock during the machining of shaft parts or retreat the tailstock to a suitable position when processing short shafts and disc parts. The operating procedure is to place the "MODE" switch at any position in the "MANU" mode; Press the "TALL STOCK INTERLOCK" button, the tailstock is released and the indicator above the button is on; Move the sliding plate to drive the tailstock to the predetermined position; Press the "TAIL STOK INTERLOCK" button again, the tailstock is locked and the indicator is off.

② Movement of the tailstock sleeve It is mainly used to tighten or loosen the workpiece when machining shaft parts. The operation method is to first place the "MODE" switch at any position in the "MANU" mode, press the "QUILL" button, the tailstock sleeve returns with the top, and the indicator is off.

（8）Clamping and loosening of the chuck. When the machine tool is operated manually or automatically, the clamping and loosening of the chuck are realized by the foot switch. The operating steps are as follows: Turn the positive and negative chuck switches in the electric box to select positive or negative chuck; Press the switch chuck for the first time to loosen it, and press the switch chuck for the second time to clamp it.

IV. Task assessment

Composition of CNC lathe and main technical specifications and parameters of CK6150e CNC horizontal lathe.

（1）Assessment form: oral examination.

（2）Assessment contents:

① Brief description of safe operation rules of the CNC lathe;

② Main technical specifications and parameters of CKA6150e CNC lathe.

（3）Assessment requirements: Simply and concisely describe the above problems and highlight the key points.

Task III   FANUC system operation panel

I. Task description

The FANUC 0i CNC operation panel consists of two parts: the controller panel（Fig. 1-3-1）and the operation panel（Fig. 1-3-2）.

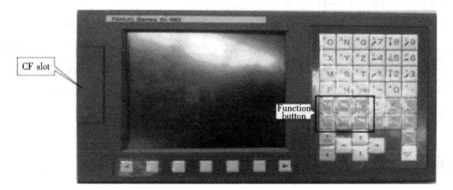

Fig. 1-3-1   FANUC 0i Controller Panel

Fig. 1-3-2   FANUC 0i Operation Panel

II. Task analysis

As shown in Fig. 1-3-1, the controller panel consists of the CRT display and the MDI keyboard. It is produced by FANUC system manufacturer. The panel operation is basically the same in the FANUC series. In terms of the operation panel, the settings of keys and knobs are different due to different manufacturers, however, their function applications are basically similar and should be flexibly used when operating on CNC lathes of different manufactures.

III. Knowledge and skills

1. Controller panel

The controller panel is the window of man-machine dialogue. See Fig. 1-3-1 for various parameters and status of the lathe, such as lathe reference point coordinates, tool starting point coordinates, command data input to the CNC system, tool offset values, alarm signals, self-diagnosis results, etc. There is a soft key operation area with 7 soft keys below the CRT display that can be used to select various CRT screens.

2.FANUC 0i operation panel

This panel (see Figure.1-3-2) is configured by the machine tool manufacturer according to the machine tool function and structure, and the arrangement and presentation of keys are different. Generally, the main functions include monitoring lights and operation keys that can be used to set and monitor the operation mode of the machine tool and CNC system. Control of the machine tool and CNC system is realized by emergency stop button, feed rate knob, spindle increase or decrease button, start/stop button, manual pulse generator, etc.

(1) System power on and off switches (see Table 1-3-1).

Table 1-3-1   System Power on and off Switches

Legends	Function description
	Name: controller on button (green button) After the button on the operation panel is triggered, the controller power supply is turned on and the CNC system is started
	Name: controller off button (red button) After the button on the operation panel is triggered, the controller power supply is turned off and the CNC system is stopped; Do not trigger this button during program operation

（2）Emergency stop（see Table 1-3-2）.

Table 1-3-2　Emergency Stop Switches

Legends	Function description
	Name: emergency stop knob（red knob） The emergency stop knob is red and is located in the middle of the left side of the machine tool operation panel. （1）In case of emergency during the operation or use of the machine tool, it is necessary to press the button immediately. The moment the button is pressed, the machine tool action can be stopped completely, and the current output to the motor is interrupted, but the machine tool will not be powered off. （2）Symptoms of pressing the emergency stop button. ① If the servo axis is running, the running axis stops moving（if the machine tool is equipped with a fourth axis and the axis is running, the fourth axis will stop running）. ② Rotating spindle stops rotation. ③ The machine tool alarm displays the following information: 1 000 EMG STOP OR OVERTRAVEL. ④ If the emergency stop knob is pressed while the cutter head is rotating, the cutter head will stop rotating immediately. ⑤ If the emergency stop knob is pressed during the tool change, the tool change action stops and enters the abnormal interruption of the tool change. （3）Pay attention to the following conditions. ① The knob can only be released after the emergency situation has been completely solved. ② At the time of stopping, all instructions and machine conditions have been deleted, so it is necessary to recheck the machining program and carry out relevant operations if there is no abnormality. （3）During automatic tool change, if press this knob, all actions stop immediately, then the cutter head may be in an uncertain position

（3）Program switch（see Table 1-3-3）.

Table 1-3-3　Program Switches

Legends	Function description
	Name: program start button Select the machining program to be executed in automatic operation mode（manual input, memory, online）and press the "program start" button to start the program
	Name: program stop button （1）In the automatic operation mode（manual input, memory, online）, after pressing the "program stop" button, each axle immediately decelerates and stops to enter the operation rest state. （2）When the button is pressed again, the machining program will continue to execute from the currently suspended single block

（4）Program protection switch（see Table 1-3-4）.

Table 1-3-4   Program Protection Switches

Legends	Function description
	Name：program protect switch （1）To prevent the program in this machine tool controller from being edited, canceled, modified and established by others, the key shall be kept by a special person. （2）Generally, set this key to "OFF" to ensure that the program is not modified or deleted. （3）Set the key to "ON" to edit, cancel or modify the program

（5）Operation mode（see Table 1-3-5）.

Table 1-3-5   Operation mode switches

Legends	Function description
	Name：edit mode "Edit mode" is displayed on the lower left corner of the machine tool display. （1）Set the program protection switch to "ON" to edit, modify, add or delete program in this mode. （2）This mode is only used for editing programs not for executing programs. （3）When a new editing program is executed, it must be switched to automatic mode（manual input, memory）before programming can be executed. （4）The controller can automatically store after the program editing is completed, and the storage action is no longer required. （5）In this mode, the machining program can be read by each personal computer
	Name：Auto mode It is displayed as "MEW" on the lower left corner of the machine tool display. （1）In this mode, press the "program start" button to execute the currently selected machining program. （2）Programs in CNC memory can be executed in this mode. （3）Feed rate in this mode refers to description of the feed rate adjustment button. （4）In this mode, the program ends when the M30 is executed and the program is restored
	Name：manual data input（MDI） （1）It is displayed as "MDI" on the lower left corner of the machine tool display. （2）In this mode, enter a single block of program command on the controller MDI panel for execution. （3）In this mode, after the command is executed, parameters can be set to determine whether the edited program needs to be eliminated. （4）Only some program segments can be entered in this mode

Legends	Function description
Hand mode	**Name: Hand mode** It is displayed as "HAND" on the lower left corner of the machine tool display. (1) In this mode, each axis can be moved with a handheld unit. (2) In this mode, the axial movement can be selected by the axial knob on the handheld unit, so as to control the movement of the selected axis. (3) In this mode, the manual movement speed of each axis can be determined by the feed rate knob on the handheld unit. (4) The rotation speed of the manual pulse generator shall not exceed 5 turns/s
JOG mode	**Name: JOG mode** It is displayed as "JOG" on the lower left corner of the machine tool display. (1) To move each axis in this mode, press each axial key and select the slow feed rate (2) In this mode, move the feed rate according to the slow feed rate. The rate adjustment range is 0-1 000mm/min. (3) When the axial key is pressed in this mode, the specified axial direction can be moved. When the key is released, the axis stops. (4) When used with the fast button, move the feed rate according to the fast feed rate. Press each axial key, the specified axial direction can be moved. When the key is released, the shaft stops
Return reference point	**Name: reference return mode** It is displayed as "REF" on the lower left corner of the machine tool display (1) In manual mode, it is used for mechanical return to zero point of the servo axis (2) After each startup of the machine tool, the reference return operation shall be performed. If the position of each axis is near the zero point, each axis needs to be manually moved away from the zero point before continuing to complete the reference return operation. (3) In this mode, after selecting the axial key requiring mechanical reference return, the zero point indicator will continue to flash until the action is completed, and then the indicator will no longer flash and be always on. (4) The mechanical reference return rate of each axis: parameter set zero return speed × rate value where the feed rate switch is located (%)

(6) Auxiliary functions (see Table 1-3-6).

Table 1-3-6　Auxiliary Function Switches

Legends	Function description
	Name: single block execution This function is only valid in automatic mode (1) When this button indicator is on, the function key of the program single block is valid. After this function is enabled, the program will be executed in a single block, and after the current single block is completed, the program will stop. The next single block of program can be executed only after the program start button is pressed, and the subsequent execution program will be so on. (2) When the key indicator is not on, the program single block execution function is invalid. The machining program will be executed until the end of the program
	Name: dry run This function is only valid in automatic mode (1) When this button indicator light is on, the *OZ* axis lock function key is valid After this function is enabled, the *F* value (cutting feed rate) command set in the program is invalid, and the movement rate of each axis is the low speed movement rate. (2) When the function is invalid, if the slow feed rate or cutting feed rate cannot change the feed rate when the program executes loop program, make a fixed feed rate according to the *F* value in the controller
	Name: skip This function is only valid in automatic mode (1) The program skip function is valid when this button indicator is on. After this function is enabled, in automatic operation, when a "/" (diagonal) symbol is specified at the beginning of the program segment, this program segment will be skipped and will not be executed. (2) The program skip function is invalid when this button indicator is off. After this function is disabled, this program segment can be executed normally even if there is a "/" (diagonal) symbol in front of the program single block
	Name: optional stop This function is only valid in automatic mode (1) The program optional stop function is valid when this button indicator is on. When this function is enabled, the program will stop in this single block if there is an M01 command during executing the program. To continue the program, press the program start key; (2) The program stop function is invalid when this button indicator is off. After this function is disabled, the program will not stop executing even if there is an M01 command in the program

Legends	Function description
	Name: mechanical lock (1) Mechanical lock function keys of all shafts are valid when this button indicator is on. After this function is enabled, CNC stops servoing any axis in manual or automatic mode The motor outputs pulses (movement command), but the command distribution is still in progress, and the absolute and relative coordinates of the corresponding axis are also updated.M, S, T, B codes will continue to be executed without mechanical lock restrictions. (2) After this function is released, it is necessary to return to the mechanical zero point again. After the zero return is correct and complete, other relevant operations can be carried out. If relevant operations are carried out without returning to zero, it will cause coordinate offset, and even abnormal phenomena such as machine collision and program running, resulting in danger
	Name: F1 This key is based on the actual configuration of the machine tool. It is a prepared space key that the operator can not operate
	Name: F2 This key is based on the actual configuration of the machine tool. It is a prepared space key that the operator can not operate
	Name: F3 Extension of working lamp. It is mainly used to control the ON and OFF of working lamps without limitation of any operating mode

(7) Spindle functions (see Table 1-3-7).

Table 1-3-7　Spindle Function Switches

Legends	Function description
	Name: spindle speed decreases (1) This button is located on the operation panel of the machine tool to reduce the programmed spindle speed S, and the actual speed = programmed S command value x spindle speed reduction rate value; (2) Use with the set spindle speed

Continue Table

Legends	Function Description
	Name: spindle speed increases. (1) This button is located on the operation panel of the machine tool to increase the programmed spindle speed S, and the actual speed = programmed S command value x spindle speed increase rate value; (2) When the program set speed exceeds the maximum spindle speed and the speed reaches more than 100% multiple, the spindle adjustment speed is equal to the maximum spindle speed; (3) Use with the set spindle speed
	Name: spindle forward (1) After the machine tool executes the S code once, select the manual operation mode, and press the spindle forward key to rotate the spindle clockwise. Spindle rotation speed = spindle speed S value previously executed $\times$ spindle adjustment knob position; (2) Operating conditions: ① It can only be used in "JOG" mode, "fast" mode and "hand" mode; ② When the spindle forward M03 command is executed in the program in automatic mode, this button indicator will be on (3) When "spindle stop" or "spindle reverse" takes effect, the indicator will be off; (4) When reverse rotation of the spindle is required, the reverse rotation operation can only be carried out after the spindle is stopped
	Name: spindle stop (1) Whether the spindle is in forward or reverse rotation, it can be stopped by pressing this key. (2) Operating conditions: ① It can only be used in "JOG" mode, "fast" mode and "hand" mode; ② Invalid for automatic operation; (3) This button indicator will light up when the spindle stops and will go out when the "spindle forward" or "spindle reverse" takes effect

Continue Table

Legends	Function Description
	Name: spindle reverse (1) After the machine tool executes the S code once, select the manual operation mode, and press the spindle reverse key to rotate the spindle counterclockwise. Spindle rotation speed = spindle speed S value previously executed × spindle adjustment knob position. (2) Operating conditions: ① It can only be used in "JOG" mode, "fast" mode and "hand" mode; ② When the spindle reverse M04 command is executed in the program in automatic mode, this button indicator will be on. (3) This button indicator will light up when the spindle reversely rotates and will go out when the "spindle forward" or "spindle stop" takes effect. (4) When forward rotation of the spindle is required, the forward rotation operation can only be carried out after the spindle is stopped

(8) Auxiliary functions (see Table 1-3-8).

Table 1-3-8　Auxiliary Functions Switches

Legends	Function description
	Name: machining cooling (1) In JOG, fast and hand modes, press this key, and the indicator light will be on, and the cutting fluid will be on; (2) Press the "RESET" key to stop coolant ejecting, the coolant stops, and the indicator light goes out; (3) Pay attention to the orientation of the coolant nozzle when the coolant is on
	Name: manual tool change In JOG, fast and hand modes, each time this key is pressed, the tool rotates one tool position along the "+" direction

(9) Axial selection key function (see Table 1-3-9).

Table 1-3-9　Axial Selection Switches

Legends	Function description
	Name: +X control button +X button: In JOG mode, press this button to move the X-axis towards the X-axis "+" direction (positive direction) of the machine tool at the speed of feed rate/fast rate, and the key indicator is on; When the button is released, the shaft stops moving towards the "+" direction and the button indicator goes out. In addition, the button also acts as the trigger key for the X-axis zero return.

Legends	Function description
	Name: −X control button −X button: In JOG mode, press this button to move the X-axis towards the X-axis "−" direction (negative direction) of the machine tool at the speed of feed rate/fast rate, and the key indicator is on; When the button is released, the shaft stops moving towards the "−" direction and the button indicator goes out
	Name: +Z control button +Z button: In JOG mode, press this button to move the Z-axis towards the Z-axis "+" direction (positive direction) of the machine tool at the speed of feed rate/fast feed rate, and the key indicator is on; When the button is released, the shaft stops moving towards the "+" direction and the button indicator goes out. In addition, the button also acts as the trigger key for the Z−axis zero return
	Name: −Z control button −Z button: In JOG mode, press this button to move the Z-axis towards the Z−axis "−" direction (negative direction) of the machine tool at the speed of feed rate/fast rate, and the key indicator is on; When the button is released, the shaft stops moving towards the "−" direction and the button indicator goes out. In addition, when the program executes the Z−axis negative movement program command, the key indicator will light up, and when the movement command is stopped, the key indicator will go out
	Name: overtravel release key (1) When the travel of each axis of the machine tool exceeds the hard limit, the machine tool will give an overtravel alarm, and the machine tool action will stop. At this time, press this button to move the machine tool overtravel axis in the opposite direction with the handheld unit in the handwheel mode. (2) It is no need to press the button for overtravel of absolute encoder machine tool
	Name: manual fast movement This function is only valid in automatic mode, and the indicator will light up when pressing the button. In this mode, actual fast feed speed = parameter setting G00 maximum speed value X rate value where fast rate switch is located (%)

(10) Feed rate and feed adjustment (see Table 1-3-10).

Table 1-3-10　Feed Rate and Feed Adjustment Switches

Legends	Function description
	Name: Feed rate and feed adjustment (1) This knob is located on the operation panel of the machine tool to control the specified G01 speed of the programming, and the actual feed rate = programmed F command value X rate value of the feed rate switch (%); (2) In manual mode, the JOG feed rate is controlled at this time, and the actual JOG feed rate = parameter set fixed value X feed rate switch value (%); (3) Use with the set shaft speed

(11) Fast rate (see Table 1-3-11).

Table 1-3-11　Fast Rate Switches

Legends	Function description
	Name: fast rate (1) This button is located on the operation panel of the machine tool to control the program set G00 speed, and the actual feed rate = parameter set G00 maximum rate value X rate value of the fast feed knob; (2) In fast mode, the manual fast feed rate is controlled at this time, and the actual fast feed rate = parameter set G00 maximum speed value × rate value of the fast feed rate switch (%). Fast movement rate can be adjusted in four gears, namely F0, 25%, 50% and 100%; (3) Use with the set axis feed speed

(12) Manual pulse (see Table 1-3-12).

Table 1-3-12　Manual Pulse Switches

Legends	Function description
	Name: Axial selector switch (1) This knob is located on the handheld unit and used in conjunction with the handwheel feed rates ×1, ×10 and ×100; (2) This knob is used for "handwheel" mode; (3) Toggle the knob to "0" position without selecting any axis, to "X" position to select X-axis, to "Y" position to select Y-axis, to "Z" position to select Z-axis and to "4" position to select 4 axis
	Name: rate selector button (1) This button is located on the handheld unit and used in conjunction with the handwheel feed rates ×1, ×10 and ×100; (2) This button is used for "handwheel" mode; (3) Toggle this button to "×1" position to select the handwheel feed rate of 0.001 mm/grid, to "×10" to select the handwheel feed rate of 0.01 mm/grid, and to "×100" to select the handwheel feed rate of 0.1 mm/grid

Continue Table

Legends	Function description
	Name: manual pulse generator（MPG） （1）This encoding disk is only valid in "handwheel" mode for controlling the feed axis direction and speed; （2）The rotation direction of the manual pulse generator is positive in the clockwise direction（i.e., the servo axis moves in the positive direction after forwarding rotation）, and negative in the counterclockwise direction（i.e., the servo axis moves in the negative direction after reverse rotation）. Warnings: （1）The rotation speed of the handwheel shall not exceed 5 turns per second. If the handwheel rotates more than 5 turns per second, the tool may not stop after the handwheel stops rotating or the distance of the tool movement does not match the scale of the handwheel rotation; （2）Handheld units shall be handled with care and shall be protected

（13）Relationship of Axis Selection, Rate Selection and Servo Axis Movement per Grid of Handwheel（see Table 1–3–13）.

Table 1-3-13　Relationship of Axis Selection, Rate Selection and Servo Axis Movement per Grid of Handwheel

Rate selection	×1	×10	×100
Metric movement	0.001 mm/grid	0.01 mm/grid	0.1 mm/grid
Imperial movement	0.000 1 in/grid	0.001 in/grid	0.01 in/grid

Note: 1 in=25.4 mm.

IV. Task implementation

1. Editing program

Toggle the MODE SELECT knob ![knob] on the operation panel to the EDIT, press ![key] on the MDI keyboard to the editing page, and after selecting an NC program, the program will be displayed on the CRT interface and can be edited.

（1）Move cursor Press PAGE ![down] or ![up] to turn pages, and press CURSOR ![down] or ![up] to move the cursor.

（2）Insert character Move the cursor to the desired position, click number or letter keys on the MDI keyboard, input the code into the input domain, press ![key], and insert the contents of the input domain behind the code where the cursor is located.

（3）Delete data in the input domain Press ![CAN] to delete the data.

（4）Delete character Place the cursor to the position where the character needs to be deleted first, and press to delete the code where the cursor is located.

（5）Search Enter letters or codes to be searched; press CURSOR to start searching behind the position where the cursor is located in the current NC program (the code may be a letter or a complete code, such as "N0010" or "M", etc). If there is a code searched in this NC program, the cursor stays at the code found; If there is no code searched, the cursor stays at the original position.

（6）Replace First move the cursor to the position of the character to be replaced, input the replaced character into the input domain through the MDI keyboard, and press to replace the code where the cursor is located with the content of the input domain.

2. Display the catalogue of NC program

Toggle the MODE SELECT knob on the operation panel to the "EDIT", press on the MDI keyboard to the editing page, and then press the soft key . The name of the NC program is displayed on the CRT interface.

Select a NC program Toggle the MODE SELECT knob on the operation panel to "EDIT" or "AUTO", press on the MDI keyboard to the editing page, and press to input the letter "O"; Press the number keys to input the search number ×××× (the search number is the program number displayed in the NC program catalogue), and press cursor to start searching. Once found, "O××××" is displayed at the program number position in the upper right corner of the screen, and the NC program is displayed on the screen.

3. Delete a NC program

Toggle the MODE SELECT knob on the operation panel to the "EDIT", press on the MDI keyboard to the editing page, and press to input the letter "O"; press the number keys to input the number XXXX of the program to be deleted; press to delete the program.

Create a new NC program Toggle the MODE SELECT knob on the operation panel to the EDIT, press on the MDI keyboard to the editing page, press to input the letter "O", and press number keys to input the program number that should not be repeated with the existing program number; press to input program; Every time a code is inputted, press and the contents in the input domain are displayed on the CRT interface, and use the Enter button to end the input of one line.

Note: The number/letter keys on the MDI keyboard are entered letters when pressed for the first time, and numbers when pressed again later. To enter letters again, the existing content in the input domain must be displayed on the CRT interface (The content in the input domain can be displayed on the CRT interface by pressing 🔲) .

4. Delete all NC programs

Toggle the MODE SELECT knob 🔲 on the operation panel to the "EDIT", press 🔲 on the MDI keyboard to the editing page, and press 🔲 to input the letter "O"; press 🔲 to input "−"; press 🔲 to input "9999"; press 🔲

5. Program simulation exercise

After the NC program is entered, the running path can be checked. First, move the cursor to the program header position, press the three buttons 🔲 at the same time, then toggle the MODE SELECT knob 🔲 on the operation panel to "AUTO", click the 🔲 command in the control panel to check running path mode, and then click the button 🔲 on the operation panel to observe the running path of the NC program. The program simulation can validate the correctness of the input program. During program simulation, pause, stop and single block execution are also effective.

V. Task assessment

Assessment of key functions and program input and validation of CNC lathe.

(1) Assessment form: oral test and operation.

(2) Assessment contents:

① Key functions of CNC lathe;

② Program input;

③ Program validation.

(3) Assessment requirements: describe the above problems in simple and concise language, highlight the key points and complete the relevant practical operations independently.

Task IV Determination of basic programming instructions and CNC lathe coordinate system

I. Task description

The essence of machining parts on CNC machine tools and ensuring the machining accuracy of the parts is to ensure the accuracy of the relative motion of workpiece and tool. Therefore, in order to control the movement of workpieces and tools during programming, it is necessary to master the two commonly used coordinate systems on the CNC machine tool, that is, machine coordinate system and workpiece coordinate system, and also the machine tool reference point, tool changing point and other contents. In addition, it is necessary to analyze and determine the machining process and parameters of the parts according to the accuracy and technical requirements of the part drawings and prepare appropriate NC machining programs with the specified NC programming code and format.

II. Task analysis

In order to simplify programming and ensure the universality of the program, a uniform standard has been established internationally for the coordinate system and direction and naming of the CNC machine tool.

At the same time, there are many kinds of CNC systems. In order to realize system compatibility, the International Organization for Standardization has formulated corresponding standards, and China has also formulated corresponding standards on the basis of the international standards. Due to the rapid development of numerical control technology and market competition, there are some incompatibilities between different systems, such as the program prepared by the FANUC 0i system can not be operated on the SIEMENS system.

Therefore, attention must be paid to the specific CNC system or machine tool in programming, and the programming should be strictly carried out in accordance with the provisions of the machine tool programming manual. However, in essence, the commands of each CNC system are set according to the actual machining requirements.

III. Knowledge and skills

1. Principles of coordinate system determination

The CNC lathe has three coordinate systems, namely mechanical coordinate system, programming coordinate system and workpiece coordinate system. The origin of the mechanical coordinate system is a fixed origin of the coordinate system set by the manufacturer when manufacturing the machine tool, also known as the mechanical zero point. It has been determined during machine tool assembly and commissioning and is the benchmark for machine tool machining. The mechanical coordinate system in use is determined by the reference point. After the machine tool system is started, the mechanical coordinate system is established by carrying out the reference return operation. Once the coordinate system is established, it will not change as long as the power supply is not cut off. The programming coordinate system is the coordinate system used in programming. Generally, the Z-axis coincides with the workpiece axis, and the X-axis is placed on the workpiece end face. The workpiece coordinate system is the coordinate system used in the machining of the machine tool, which shall be consistent with the programming coordinate system. Whether the programming coordinate system can be in line with the workpiece coordinate system is the key to operation.

The coordinate system on the CNC machine tool adopts the Cartesian coordinate system (right-hand and right-angled coordinate), as shown in Fig. 1-4-1. The thumb, index finger and middle finger of the right hand remain perpendicular to each other, with the thumb in the positive direction of the X-axis, the index finger in the positive direction of the Y-axis, and the middle finger in the positive direction of the Z-axis.

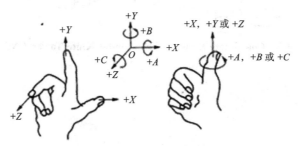

Fig. 1-4-1 Cartesian Coordinate System (Right-hand and Right-angled Coordinate)

2. Determination of movement direction

The positive direction of a part of the machine tool is the direction of increasing the distance between the workpiece and the tool, as shown in Fig. 1-4-2.

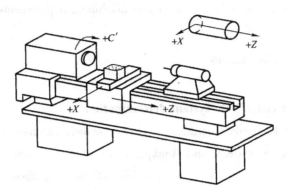

Fig. 1-4-2　Direction of Motion

(1) The Z-axis coincides with the spindle axis. Set the direction of the Z-axis away from the workpiece and the tool moving to the tailstock as the positive direction (i.e. increasing the distance between the workpiece and the tool), and the direction of moving the tool to the chuck as the negative direction.

(2) The X-axis is perpendicular to the Z-axis, and the positive direction of the X-coordinate is the direction in which the tool leaves the rotation centerline and vice versa

3. Relevant points of CNC lathe

1) Origin of machine tool

According to the coordinate system specification of the CNC lathe, as shown in Fig. 1-4-3, the spindle transmitting the cutting force is usually defined as the Z-axis. The origin of the CNC lathe is generally set at the intersection of the spindle rotation centerline and the rear end face of the chuck, as the O point shown in Fig. 1-4-4.

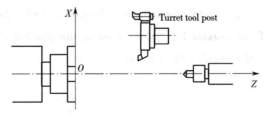

Fig. 1-4-3　Coordinate System of Horizontal Knife Tower CNC Lathe

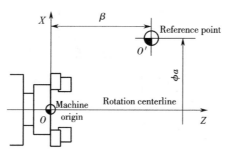

Fig. 1-4-4 Origin of Machine Tool

2) Reference point of machine tool

The machine tool reference point is also a fixed point on the machine tool, which is the limit position for limiting the movement of the tool post with mechanical stoppers or electrical devices. It is mainly used to position the machine coordinate system.

Because if the system sets the current position to (0,0) no matter where the tool post stays after each startup, which will inevitably result in inconsistency of the benchmark, therefore, the first step of each startup is to carry out reference return operation (some are called zero return), that is, to determine the origin (0,0) of the machine tool coordinate system by determining the reference point. Reference point return is to make the tool post automatically return to the fixed point of the machine tool according to command. This function is also used to check the correctness of the coordinate system and establish the machine tool coordinate system during the machining process, so as to ensure accurate control of the machining dimensions. As shown in Fig. 1-4-4, 0 point is commonly used as a tool change point, $\phi\alpha$ and β are the limit travel distances in the X and Z directions of the machine tool, i.e. the theoretical machining range of the machine tool.

When the machine tool post returns to the reference point, the coordinate value of the tool post benchmark in the machine coordinate system is a set of determined values. When the machine tool is powered on and before returning to the reference point, no matter where the tool post is, the Z and X coordinate values displayed on the CRT are 0. Only after the operation of returning to the reference point is completed, the CRT immediately shows the coordinate values of the tool post benchmark in the machine coordinate system, that is, the machine tool coordinate system is established.

3) Origin of workpiece coordinate system

During NC programming, first set a point on the part drawing as the origin of the programming coordinate according to the shape characteristics and dimensions of the machined part, which is called the program origin, as O point shown in Fig. 1-4-5.

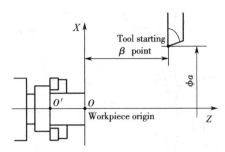

Fig. 1-4-5   Workpiece Coordinate System

Route arrangement, numerical processing and programming can only be carried out when all geometric elements on the part have determined positions. At the same time, it also determines the placement direction of the part on the machine tool during NC machining. Theoretically, the origin of the workpiece coordinate system can be selected at any point on the workpiece, but this may cause complicated calculation problems and increase the difficulties of programming. For the convenience of calculation and simplification of programming, the origin of the workpiece coordinate system is usually selected on the rotation centerline of the workpiece, and the specific location can be considered to be set on the left end face (or right end face) of the workpiece, so that the programming benchmark coincides with the design and positioning benchmarks as far as possible.

（1）Set the origin of the coordinate system on the chuck surface（Fig. 1-4-6）.

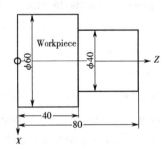

Coordinates and dimensions on machining drawings

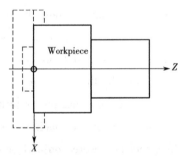

Coordinates of CNC commands on lathes（coincide with programming coordinate system）

Fig. 1-4-6   Workpiece Origin at Left End

（2）Set the origin of the coordinate system on the end face of the part（Fig. 1-4-7）.

When machining parts on the CNC machine tool, the relative movement of the tool and the workpiece must be carried out in the determined coordinate system. The programmer must be familiar with the machine coordinate system. The purpose of specifying the coordinate axis and motion direction of the CNC machine tool is to accurately describe the motion of machine tool, simplify the programming method, and make the programming interchangeable.

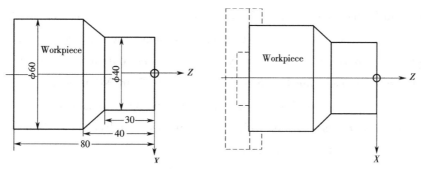

Fig. 1-4-7 Workpiece Origin at Right End

The machine coordinate system is the only benchmark of the machine tool, so it is necessary to clarify the position of the program origin in the machine tool coordinate system. This is usually done in the next tool setting process. The essence of the tool setting is to determine the position of the origin of the workpiece coordinate system in the only machine coordinate system. Tool setting is the main operation and important skill in NC machining. The accuracy of the tool setting determines the machining accuracy of the part, and the tool setting efficiency also directly affects the NC machining efficiency.

4) Tool changing point

When tool change is required during machining of the CNC lathe, appropriate tool changing points shall be considered during programming. The so-called tool changing point refers to the position of changing tools by the tool changer. When the workpiece coordinates are determined on the CNC lathe, the tool changing point may be a fixed point or any point relative to the origin of the workpiece. The tool changing point shall be outside the workpiece or the fixture, subject to the fact that the tool changer does not touch the workpiece and other parts when the tool is changed.

4. Programming method

(1) Absolute value programming: A method to calculate the absolute value coordinate dimension for programming according to the preset program origin. The absolute value coordinates are represented by addresses X and Z in the FANUC system, of which the X refers to the diameter value.

(2) Incremental value programming: a programming method that represents the position based on the increment of coordinate values from the previous position. Addresses U and W represent incremental coordinates, U represents diameter increments

(3) Mixed programming: a method by mixing absolute and incremental values for programming.

5. Function commands of FANUC CNC system

During the programming of CNC machine tools, various actions in the machining process, such as the opening, stopping and reversing of machine tool spindle, the feed direction of tool, the opening and closing of coolant, shall be specified in the form of commands. Such commands are called function commands. The function commands used in the NC program mainly include preparatory function G command, auxiliary function M, feed function F, spindle speed function S and tool function T. In NC programming, various G and M commands are used to describe the machining and motion characteristics. At present, the ISO-1056-l975E standard is widely used internationally, and China formulated GB/T38267—2019 standard based on it.

1）Preparatory function G command of the FANUC CNC system

The preparatory function command, also known as G command or G code, is a command to establish the working mode of the machine tool or control CNC system. Such command shall be specified in advance before the interpolation operation of the CNC device so as to prepare for interpolation, tool compensation, fixed loop, etc. The G command is composed of the letter G and two numbers. Table 1-4-1 shows the list of common commands for the FANUC CNC lathe.

Table 1-4-1 Common G Commands for FANUC CNC Lathe

Code G	Group	Function	Code G	Group	Function
*G00	01	Positioning（rapid motion）	G55	14	Select workpiece coordinate system 2
G01		Straight cutting	G56		Select workpiece coordinate system 3
G02		Circular interpolation（CW, clockwise）	G57		Select workpiece coordinate system 4
G03		Circular Interpolation（CCW, counterclockwise）	G58		Select workpiece coordinate system 5
G04	00	Pause	G59		Select workpiece coordinate system 6
G20	06	Imperial input	G70	00	Finishing loop
G21		Metric input	G71		Rough turning of inner and outer diameters
G27	00	Check reference point return	G72		Step rough turning
G28		Reference point return	G73		Repeated fixed shape cycles
G29		Return from reference point	G74		Feed drilling in Z direction
G30		Return to the second reference point	G75		X-direction groove
G32	01	Thread cutting	G76		Thread cutting loop

Code G	Group	Function	Code G	Group	Function
*G40		Cancel offset of tool tip radius	G90		Cutting loop (internal and external diameters)
G41	07	Radius offset of tool tip (left)	G92	01	Thread cutting loop
G42		Radius offset of cutter tip (right)	G94		Cutting loop (step)
G50		Maximum spindle speed setting (coordinate system setting)	G96		Constant linear speed control
G52	00	Set local coordinate system	*G97	02	Cancel constant linear speed control
G53		Select machine coordinate system	G98		Specify the movement per minute
*G54	14	Select workpiece coordinate system 1	*G99	05	Specify the movement per second

Note: 1. The command marked with "*" is the command that has been set when it is started;

2. The "00 group" G codes are non-modal commands, and can only work in one program segment;

3. Several G commands from different groups can be used in one program segment, and the last G code is valid if more than one G command from the same group is used.

The G commands can be divided into three kinds in terms of function.

(1) is the machining mode G codes and the machine tool has corresponding actions when executing such G codes. The corresponding coordinate value must be specified programming format, such as "G01 X60.Z0;".

(2) is the function selection G codes, which are equivalent to the function start and end selections, and no address character is specified during programming. The internal default functions of the CNC machine tool after power on generally include setting absolute coordinate mode programming, using metric length unit dimensions, canceling tool compensation, spindle and cutting fluid pump shutdown and other states as its initial state.

(3) is the parameter setting or calling G codes, such as the G50 coordinate setting command which only changes system coordinate parameters during execution; When the G54 is executed, only the system parameters are called and the machine tool will not act.

2) Auxiliary function M command of FANUC CNC system

The auxiliary function command is also called the M command or M code. Such commands are used to specify the auxiliary functions of the machine tool or system, such as the coolant on/off, spindle forward/reverse rotation and program end, etc. In the same program segment, if there are two or more auxiliary function commands, read the later ones.

The M command is composed of the letter M and two numbers. See Table 1-4-2 for common auxiliary function commands of FANUC system.

Table 1-4-2   Common M Commands for FANUC CNC Lathe

Function M	Meaning	Function M	Meaning
M00	Program stop	M08	Coolant on
M01	Optional stop	M09	Coolant off
M02	Program end	M30	Program end and return to starting position
M03	Spindle rotates clockwise	M98	Call for subprogram
M04	Spindle rotates counterclockwise	M99	Return from subprogram
M05	Turn spindle off		

3) Other common functions of FANUC system

In addition to G and M commands, a standard program shall also have F, S and T functions when programming.

(1) F function, also called feed function, is used to specify the feed speed of the tool. It is composed of the letter F and numbers in programming, which can set the feed unit with the feed G98 per minute and the feed G99 command per rotation.

(2) S function, also known as the spindle speed function, is used to specify the rotation speed of the spindle. It is composed of the letter S and numbers in programming.

(3) T function, also known as the tool function, is used to specify the tool number and the tool compensation number. It is composed of the letter T and numbers in programming, and the numbers can be two or four digits according to the tool quantity of the machine tool.

6. NC programming format of FANUC system

1) Basic methods for NC program preparation

(1) Contents and steps of NC program preparation.

As shown in Fig. 1-4-8, programming mainly includes:

① Analysis of part drawings and development of machining scheme. Analyze the part drawings to define the contents and requirements of machining; determine the machining scheme; select appropriate CNC machine tool; select or design tools and fixtures; determine a reasonable travel path and select proper cutting amount, etc. This requires that the programmer can analyze the technical characteristics, geometry, dimensions and machining requirements of the part drawings, and determine the machining methods and routes in combination with the basic knowledge of the CNC machine tool, such as the specifications, performance and functions.

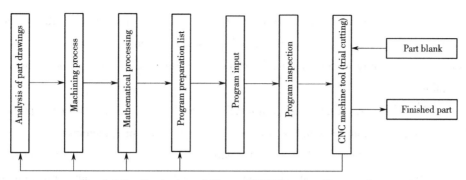

Fig. 1-4-8　Contents and Steps of NC Program Preparation

② Mathematical processing. After the machining process scheme is developed, it is necessary to calculate the central motion trajectory of the tool according to the geometric dimensions of the part, the machining route, etc. to obtain the tool path data. The CNC system generally has the functions of linear interpolation and circular interpolation. For machining simple plane parts consisting of arcs and straight lines, the programming requirements can be met only by calculating the coordinate values of the intersection points or tangent points of adjacent geometric elements on the part profile and obtaining the center coordinate values of starting points, end points and arc of each geometric element. When the geometry of the part is inconsistent with the interpolation of the control system, a more complex numerical calculation is required, which generally uses computer-aided calculation, or it is difficult to complete.

③ Prepare parts machining programs. The part machining program can be prepared after the above operation and numerical calculation are completed. The programmer uses program commands of the CNC system and prepares the machining program section by section according to the specified program format. Only when the programmer is familiar with the functions, program commands and codes of the CNC machine tool can the programmer prepare a correct machining program.

④ Program inspection. By inputting the programmed machining program into the NC system, the machining work of the CNC machine tool can be controlled. Generally, programs should be inspected before formal operation. In general, the machine tool can be idled to check the correctness of the machine tool's action and motion trajectory so as to inspect the program. On the CNC machine tool with graphic simulation display function, the program can be checked by displaying the feed path or simulating the cutting process of the tool on the workpiece. For parts with complex shapes and high requirements, the program can also be inspected by trial cutting with easily cut materials such as aluminum parts, plastic or paraffin. By checking the test part, it can be confirmed that whether the program is correct and that the machining accuracy meets the requirements. If the same materials as the machined parts can be used for trial cutting, it can better reflect the actual machining effect. When it is found that the machined parts do not meet the machining technical requirements, the program can be modified or measures such as dimension compensation can be taken.

(2) Method for NC program preparation.There are two main methods for the preparation of NC machining program: manual programming and automatic programming.

① Manual programming. Generally, for parts with less complex geometry, the required machining program is not long, the calculation is relatively simple, and manual programming is appropriate.

Characteristics of manual programming: it takes a long time, is prone to errors, and is not competent for the programming of parts with complex shapes. According to foreign data statistics, when manual programming is adopted, the ratio of the preparation time of a program to the actual time of its operation and machining on the machine tool is about 30: 1 on average, while 20% to 30% of the reasons why the NC machine tool can not be started are due to the difficult preparation of the machining program and the programming time is long.

② Computer automatic programming. Automatic programming means that during programming, except that the part drawings are analyzed and the machining scheme is developed manually, the rest of the work is completed with the aid of computer software.

When automatic computer programming is adopted, mathematical processing, programming, inspection program and other works are automatically completed by the computer. Since the computer can automatically draw the movement trajectory of the tool center, the programmer can check the program in time for correctness, and can modify it in time if necessary to obtain the correct program. In addition, computer automatic programming replaces programmers to complete cumbersome numerical calculations, which can improve programming efficiency by dozens or even hundreds of times, solving the programming problems of many complex parts that cannot be solved by manual programming. Therefore, the characteristics of automatic programming are that it has high programming efficiency and can solve the programming problems of complex shape parts.

2) NC programming format of FANUC CNC system

(1) Format of program.The preparation of the machining program is to write control commands according to the actual sequence of machine actions and tool routes. Commands arranged in sequence are called program segments. For continuous machining, many program segments are required, and the set of these segments is called program. The numbers added to identify each program segment are called sequence numbers, while the numbers added to identify each program are called program numbers. A complete program is generally

composed of program number, program content and program end. The format is as follows:

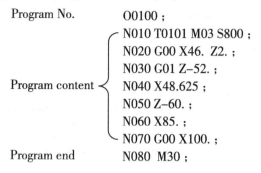

Program No.　　　　　O0100 ;

Program content
- N010 T0101 M03 S800 ;
- N020 G00 X46. Z2. ;
- N030 G01 Z−52. ;
- N040 X48.625 ;
- N050 Z−60. ;
- N060 X85. ;
- N070 G00 X100. ;

Program end　　　　　N080　M30 ;

The program number is used as the start identification of the machining program. Each workpiece machining program has its own special program number. Different CNC systems have different program numbers and address codes. Symbols such as %, P and O are commonly used. Programming must be specified according to the provisions of the system manual, such as writing in the form of %8, P10 and O0001, otherwise, the system cannot recognize them. The program content consists of several program segments of machining sequence, various motion paths of tools and various auxiliary actions. The end symbol indicates the end of the machining program, for example, M02 in the FANUC system; if the program is required to return to the beginning of the program, the M30 command is required.

Each coordinate value in the program segment shall be entered with at least one decimal place, and "; " shall be added at the end of each program segment to indicate the end of this segment.

(2) Format of program segment.A program segment defines a command line to be executed by the CNC device. The format of the program segment defines the syntax of the function words in each program segment, and its structure is shown in Fig. 1-4-9 below.

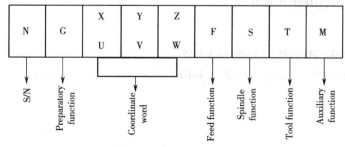

Fig. 1-4-9   Format of Program Segments

(3) Format of program command word.A command word consists of an address character (command character) and data with a symbol (e.g. a word defining dimension) or without a symbol (e.g. preparatory function word G code) . Different command characters

and subsequent data in the program define the meaning of each command character, and the common addresses contained in the NC program segment are shown in Table 1-4-3.

Table 1-4-3　List of Command Characters

Function	Command character		Meaning
Program number	O		Program numbers（0~9 999）
Sequence number of program segment	N		Sequence number of program segment （NO~N...）
Preparatory function	G		Command action mode（straight, arc, etc.）
Dimension word	X, Y, Z, D, V, W, A, B, C		Movement of coordinate axis
	R		Parameters of arc radius and fixed loop
	I, J, K		Center coordinates
Feed function	F		Feed speed determination
Spindle function	S		Spindle speed determination
Tool function	T		Tool number selection
Auxiliary function	M		Machine on/off and related control
Pause	P, X		Pause time determination
Subprogram number determination	P		Subprogram number determination
Repeat times	L		Repeat times of subprogram
Parameter	P, Q, R, U, W, I, K, C, A		Parameters of turning compound cycles
Chamfer control	C, R		Automatic chamfer parameters

IV. Task implementation

1. Machining process of typical parts

Machine the part shown in Fig. 1-4-10, blank ϕ45 mm long bar, requiring one-time clamping.

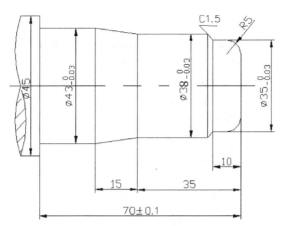

Fig. 1-4-10 Shaft Part

1) Process analysis

(1) The appearance of the part is complex and the cylindrical, taper, convex arc and chamfer need to be machined.

(2) Select tool according to the graphic shape.

T01 cylindrical rough tool: large machining allowance, requiring no interference of the tool minor cutting edge angle.

T02 cylindrical finishing tool: diamond blade, 0.4 tool tip arc, tool minor cutting edge angle >35 °.

(3) Coordinate calculation: According to the selected commands, if this part is programmed with G01 and G02 commands, the rough machining route is complex, especially the calculation and programming at the arc; it is suitable to use G71 command, and the coordinate points of finish turning outer shape are obtained according to the graphics during machining, and the programming is processed at one time.

2) Processing technology and programming route of FANUC CNC system

(1) Flatten end face of #1 tool.

(2) Rough machining the outer shape of the #1 tool with the G71 command.

(3) Finishing machining the outer shape of the #2 tool with the G70 command.

3) Reference Programs for FANUC CNC System

Table 1-4-4 Programs for Shaft Parts

01401;	Program name;
N1;	Program segment number (rough machining segment);
G99 T0101 M03 S600;	Change #1 cylindrical tool, spindle forward rotation, speed 600 r/min;
G00 X100. Z100. M08;	Fast motion, intermediate safety point, coolant on;
G00 X47.Z2.;	Starting point of loop;
G71 U1.5 R0.5;	Compound loop turning of outer shape, X-direction back cutting depth 1.5 mm, retract amount 0.5 mm;

Continue Table

G71 P10 Q20 U0.5 W0.02 F0.2;	Finishing machining program segment N10-N20, X-direction allowance 0.5 m, Z-direction 0.02 mm;
N10 G00 X0.	Finishing machining the first segment;
G01 Z0 F0.1;	Flatten end face;
G01 X25. ;	Starting point of arc;
G03 X35.Z−5.R5;	Machine convex arc;
G01 Z−10.;	Machine ϕ35 mm cylindrical;
X38. C1.5;	Chamfer;
Z−35.;	Machine ϕ38 mm cylindrical;
X43. W−15.;	Machine taper;
Z−70.;	Machine ϕ43 mm cylindrical;
N20 G00 G40 X47.;	Retract tool;
G00 X100. Z100.;	Return to tool changing point;
M05;	Spindle stop;
M09;	Coolant off;
M00;	Program stop;
N2;	Finishing machining;
G99 M03 S800 T0202;	Change #2 cylindrical finishing tool;
G00 X47. Z2.;	Starting point of loop;
G70 P10 Q20 F0.;	Finishing machining of outer shape;
G00 X100. Z100.;	Return to tool changing point;
M05;	Spindle stop;
M09;	Coolant off;
M30;	Program stop.

V. Task assessment

Assessment of CNC lathe program preparation and inspection.

（1）Assessment form: operation.

（2）Assessment contents:

① Program preparation;

② Program input;

③ Program validation.

（3）Assessment requirement: independently complete relevant actual operations.

Item 2

Programming and Machining of

Shaft Workpiece

[Item description]

This item is one of the basic contents of NC machining that will involve more knowledge on NC programming and machining. For example, it is necessary to master the clamping methods of shaft parts and tools, fill in the machining tool card and process card, complete the programming and machining of shaft parts, and get familiar with the inspection methods of shaft parts. Furthermore, theory should be contacted with practice, specifically carrying out simulation machining with NC simulation software in the programming classroom, entering the training workshop site for machine tool processing, and completing the machining tasks in groups or separately.

[Learning objectives]

（1）Understand the structural characteristics of shaft parts.

（2）NC turning process analysis of shaft parts.

（3）Master the machining programs for shaft parts prepared with commands G00, G01, G02, G03, G90, G71, G73, G70, G32 and G92.

（4）Correctly select and use the tools and cutting amount commonly used for shaft parts.

（5）Be able to operate the FANUC 0i CNC lathe to complete part machining.

Task I   Programming and machining of step shaft part

I. Drawings and technical requirements

Among many machined parts, shaft parts are undoubtedly one of the most important and commonly used mechanical elements in the machinery industry. This task focuses on the design, processing technology and machining of the step shaft. The step shaft shown in Fig. 2-1-1 is the task to be completed this time, and the blank is ϕ44 mm×73 mm.

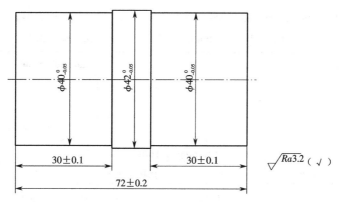

Fig. 2-1-1 Step Shaft

II. Analysis of drawings and process arrangement

1. Analysis of drawings

Fig. 2-1-1 shows the step shaft with a simple shape and little change in structural size. There are 3 step surfaces. Among the radial dimensions, the accuracy tolerance of $\phi40$ mm and $\phi42$ mm is 0.05 mm, which is not high. There is a length tolerance requirement for the $\phi40$ mm cylindrical section in the axial dimension, and the surface roughness value is not more than $Ra3.2$ μm.

2. Process arrangement

1) Determine the clamping scheme of the workpiece

According to the shape of the blank, the scroll chuck can be selected for clamping, and the program origin is set at the intersection of the right end face of the workpiece and the axis. This part can only be machined by secondary clamping. For the first time, the left end of the workpiece is clamped, and the cylindrical turning is $\phi42$ mm and $\phi40$ mm. For the second time, the $\phi40$ mm finishing cylindrical turning is used as the positioning benchmark, the end face is flattened, the total length is determined, and then the $\phi40$ mm cylindrical is machined.

2) Determine the machining route

(1) Flatten end face and determine benchmark;

(2) Rough and finishing turning $\phi42$ mm and $\phi40$ mm cylindrical;

(3) Workpiece turning around and clamping $\phi40$ mm cylindrical;

(4) Flatten end face (determine the total length);

(5) Rough and finishing turning $\phi40$ mm cylindrical.

3）Fill in the machining tool card and process card（see Table 2-1-1）

Table 2-1-1　Machining Tool Card and Process Card

Part drawing No.	2-1-1	CNC lathe process card		Machine model	CKA6150
Part name	Step shaft			Machine tool number	
Table of turning tools				Table of measuring tools	
Tool number	Tool compensation number	Name of turning tools	Turning tool parameters	Measuring tool name	Specification/mm
T01	01	90° cylindrical turning tool	Type C blade	Vernier caliper Micrometer	0-150/0.02 25-50/0.01
Working procedure	Process content	Cutting amount			Processing property
		S/（r/min）	F/（mm/r）	a_p/mm	
CNC lathe	Cylindrical turning and radial turning for determining reference	500	—	1	Manual
1	Machining ϕ42 mm cylindrical	600-1 000	0.1-0.2	0.5-3	Automatic
2	Machining ϕ40 mm cylindrical	600-1 000	0.1-0.2	0.5-3	Automatic
CNC lathe	Turn around to clamp ϕ40 cylindrical, flatten end face, keep the total length	500	—	1	Manual
1	Machining ϕ40 mm cylindrical	600-1 000	0.1-0.2	0.5-3	Automatic

III. Program preparation and machining

1. G00 Fast point positioning command

The G00 command enables the tool to move quickly to the specified coordinate point position for idle travel movement before the tool is machined or quick retraction after machining. The command enables the tool to move quickly to the specified point to improve the machining efficiency and can not be cut.

Format of command：G00 X（U）_Z（W）_；

Note：

（1）Absolute value programming：G00 X_ Z_ represents the coordinate value of the end position relative to the workpiece origin. The axial moving direction is determined by the coordinate value of Z. When the tool is fed or retracted in the radial direction that doesn't pass over the axis，the value is positive. For example，two axes move G00 X80.Z10. at the same time，one axis move G00 X50.or G00 Z−10.

（2）Incremental value programming：G00 U_W_；U_W_represents the distance and direction of the tool from the current point to the end point; U represents the moving amount in the diameter direction, that is, the difference between the large and small diameters, W represents the moving length, and the moving directions of U and W are determined by positive and negative signs. The starting point coordinate value for calculating the U and W moving distances is the end point value for executing the movement command of the previous program segment. Mixed programming can also be used in the same movement command. For example, G00 U20. W30., G00 U-5. Z40. or G00 X80. W40. .

(3) When programming with G00, it can also be specified as G0.

For example, to quickly move the tool to the specified position as shown in Fig. 2-1-2, write the program segment with G00:

Absolute mode programming: G00 X50.0 Z6.0;

Incremental mode programming: G00 U-70.0 W-84.0;

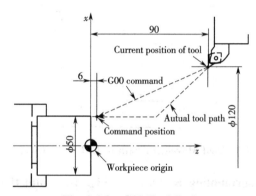

Fig. 2-1-2 G00 Feed Path

2. G01 feed cutting command

G01, also known as linear interpolation, commands the tool to move to the specified position at the specified feed speed. When the spindle rotates, it can be used to cut and machine the workpiece at a certain speed.

Format of command: G01 X (U) _Z (W) _;

Note:

(1) Absolute value programming: G01 X_ Z_ represents the coordinate value of the end position relative to the workpiece origin. The axial movement direction is determined by the coordinate value of Z. When the tool is fed or retracted in the radial direction that doesn't pass over the axis, the value is positive. For example, two axes move G01 X80. Z10. at the same time, one axis moves G01 X50. or G01 Z-10.;

(2) Incremental value programming: G01 U_W_; U_W_represents the distance and direction of the tool from the current point to the end point; U represents the moving amount in the diameter direction, that is, the difference between the large and small diameters, W

represents the moving length, and the moving directions of U and W are determined by positive and negative signs. The starting point coordinate value for calculating the U and W moving distances is the end point value for executing the movement command of the previous program segment. Mixed programming can also be used in the same movement command. For example, G01 U20. W30., G01 U−5. Z40. or G01 X80. W40. .

（3）When the machine tool executes the G01 command, it must have F command in this program segment or before this program segment, otherwise the feed speed is considered zero. Unit: mm/r.

（4）Use G01 to command turning, and the tool movement path is shown as in Fig. 2-1-3.

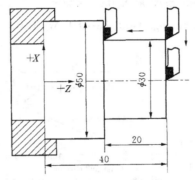

Fig. 2-1-3　Tool Movement Path of G01 Cylindrical Turning

3. Example of programming is shown in Fig. 2-1-4 and the programs are shown in Table 2-1-2.

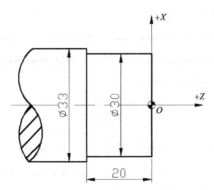

Fig. 2-1-4　G01 Cylindrical Turning

Table 2-1-2   Programs for G01 Cylindrical Turning

Program content	Program description
02001;	Program number;
N010 G99 M03 S600 T0101;	Feed rate mm/r, spindle forward rotation 600 r/min, select #1 tool;
N020 G00 X100.Z100.;	Tool changing point;
N030 G00 X30.5 Z2.;	Tool alignment point;
N040 G01 X30.5 Z−20. F0.2;	Rough turning ϕ30 cylindrical, leave 0.5 mm finishing allowance
N050 G01 X34. Z−20.;	X−direction tool retraction;
N060 G00 X34.Z2. S1000;	Z−direction tool retraction, give finishing rotation 1 000 r/min ;
N070 G01 X30. Z2. F0.1;	Finishing turning ϕ30 cylindrical starting point;
N080 G01 X30. Z−20.;	Finishing turning ϕ30 cylindrical end point;
N090 G01 X34. Z−20.;	X direction tool retraction;
N100 G00 X100. Z100.;	Return to tool changing point;
N110 M05;	Spindle stop;
N120 M30;	Return to program start after program end

4. Single loop command (G90) for shaft part machining and programming

1) Programming format:

G90 X (U) _Z (W)_F_;

Notes:

X, Z—end point coordinates for cylindrical cutting;

U, W—coordinate components of the end point of cylindrical cutting relative to the starting point of the loop;

F—feed speed.

2) Single fixed loop cutting

As shown in Fig. 2-1-5. The tool starts to loop in a form of a rectangle from the starting point and finally returns to the starting point. The dotted line in the figure indicates rapid movement according to R and the solid line indicates movement according to the feed speed specified by F.

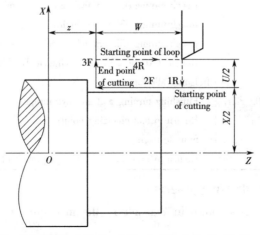

Fig. 2-1-5   Feed Path of Single Fixed Cycle Cutting

Feed path: Single fixed loop allows a series of continuous machining actions, such as "feeding—cutting—retracting—returning", to be completed with a loop command, thus simplifying the program.

Attention:

When using the loop cutting command, the tool must be positioned to the loop starting point first, then the loop cutting command is executed, and after completing one loop, the tool still returns to the starting point of the loop.

3) Example of programming is shown in Fig. 2-1-6 and the programs are shown in Table 2-2-1

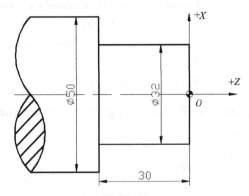

Fig. 2-1-6 G90 Cylindrical Turning

Table 2-1-3 Programs for G90 Cylindrical Turning

Program content	Program description
02201;	Program number;
N010 G99 M03 S600 T0101;	Feed rate mm/r, spindle forward rotation 600 r/min, select #1 tool;
N020 G00 X100. Z100.;	Tool changing point;
N030 G00 X55.0 Z5.0;	Starting point of loop;
N040 G90 X46.0 Z–30. F0.2;	Rough turning ϕ32 mm cylindrical, first tool, cutting depth 4 mm;
N050 X42.;	Rough turning ϕ32 mm cylindrical, second tool, cutting depth 4 mm;
N060 X38.;	Rough turning ϕ32 mm cylindrical, third tool, cutting depth 4 mm;
N070 X34.;	Rough turning ϕ32 mm cylindrical, fourth tool, cutting depth 4 mm;
N080 X32.4;	Rough turning ϕ32 mm cylindrical, fifth tool, leaving 0.4 mm finishing allowance;
N090 X32.0 Z–30.0 S1000 F0.1;	Finishing turning ϕ32 mm cylindrical;
M100 G00 X100. Z100.;	Return to tool changing point;
N110 M05;	Spindle stop;
N120 M30;	Return to program start after program end

5. Fig. 2-1-1 Part drawing program

According to the parts shown in Fig. 2-1-1, the machining route of the workpiece is analyzed, and the clamping scheme during machining, together with the tool and cutting amount, is determined. It is divided into two parts according to the machining content, and two corresponding programs are prepared to complete the machining.

Table 2-1-4 shows the programs for machining the right end of the step shaft.

Table 2-1-5 shows the programs for machining the left end of the step shaft.

Table 2-1-4 Programs for Machining the Right End of the Step Shaft

Program content	Program description
02202;	Program number;
N010 G99 M03 S600 T0101;	Feed rate mm/r, spindle forward rotation 600 r/min, select #1 tool;
N020 G00 X100.Z100.;	Tool changing point;
N030 G00 X45. Z2.;	Starting point of loop;
N040 G90 X42.5 Z−42. F0.2;	Rough turning $\phi 42$ mm cylindrical, leave 0.5 mm finishing allowance;
N050 X40.5 Z−30.;	Rough turning $\phi 40$ mm cylindrical, leave 0.5 mm finishing allowance;
N060 X40. Z−30. S1000 F0.1;	Finishing turning $\phi 40$ mm cylindrical, give finishing rotation and feed rate;
N070 X42. Z−42.;	Finishing turning $\phi 42$ mm cylindrical;
N080 G00 X100. Z100.;	Return to tool changing point;
N090 M05;	Spindle stop;
N100 M30;	Return to program start after program end

Table 2-1-5 Programs for Machining the Left End of the Step Shaft

Program content	Program description
02203;	Program number;
N010 G99 M03 S600 T0101;	Feed rate mm/r, spindle forward rotation 600 r/min, select #1 tool;
N020 G00 X100.Z100.;	Tool changing point;
N030 G00 X45. Z2.;	Starting point of loop;
N040 G90 X40.5 Z−30. F0.2;	Rough turning $\phi 40$ mm cylindrical, leave 0.5 mm finishing allowance;
N050 X40. Z−30. S1000 F0.1;	Finishing turning $\phi 40$ mm cylindrical, give finishing rotation and feed rate;
N080 G00 X100. Z100.;	Return to tool changing point;
N090 M05;	Spindle stop;
N100 M30;	Return to program start after program end

6. Parts machining

(1) The teacher demonstrates the whole preparation and machining process of the workpiece.

(2) Explain and demonstrate the guarantee method of dimensional tolerance.

(3) The students complete the machining of the workpiece in groups, and the teacher

gives tour guidance to find and correct the problems of students arising from machining in time.

IV. Task assessment

Task evaluation as per Table 2-1-6.

Table 2-1-6 Comprehensive Scoring of Step Shaft

Assessment item		Assessment requirement	Assigned score	Scoring criteria	Test result	Score	Remark
1	Outer circle	$\phi 42_{-0.05}^{0}$ mm	10	Deduct 5 points for each 0.01 mm out of tolerance			
2		$\phi 40_{-0.05}^{0}$ mm	15×2	Deduct 5 points for each 0.01 mm out of tolerance			
3	End face	Ra3.2 μm（two places）	5×2	Full deduction for out-of-tolerance			
4	Chamfering	Four places	5×4	Full deduction for out-of-tolerance			
5	Length	30 mm, 30 mm，72 mm	5×3	Full deduction for out-of-tolerance			
6	Surface	Ra1.6 μm （three places）	5×3	Full deduction for out-of-tolerance			
7	Process，program	Provisions on process and program		Deduct 1-5 points for violation of regulations			
8	Standardized operation	Regulations on CNC lathe standard operation		Deduct 1-5 points for violation of regulations			
9	Safe and civilized production	Relevant provisions on safe and civilized production		Deduct 1-5 points for violation of regulations			
Summary and feedback during students task implementation:							
Teacher's review:							

Task II   Programming and machining of arc and taper parts

I. Drawings and technical requirements

The part drawing of this task consists of concave and convex arcs, tapered surfaces, cylindrical surfaces, etc. It is relatively complex, and it is difficult to select commands, prepare programs and design the turning route. The main reason is that the allowance removed at one time during the semi-finishing of concave arc is large, which is easy to cause accidents such as turning tool. To solve these problems, this task will introduce the use of G73—pattern repeating rough turning loop.

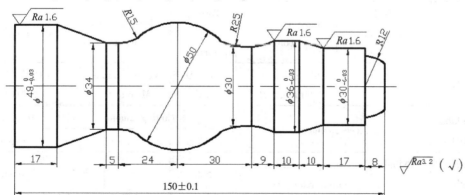

Fig. 2-2-1   Arc and Taper Parts

II. Analysis of drawings and arrangement

1. Analysis of drawings

Fig. 2-2-1 shows the arc and taper parts, the blank material is $\phi 50$ mm $\times$ 220 mm, and it is required to machine a single piece according to the drawing. The part drawing of this task consists of concave and convex arcs, tapered surfaces, cylindrical surfaces, etc. It is relatively complex，and it is difficult to select commands，prepare programs and design the turning route. The main reason is that the allowance removed at one time during the semi-finishing of concave arc is large，which is easy to cause accidents such as turning tool. To solve these problems，this task will introduce the use of G73—pattern repeating rough turning loop.

2. Process arrangement

1）Determine the clamping scheme of the workpiece

According to the shape of the blank，the scroll chuck can be selected for clamping，and the program origin is set at the intersection of the right end face of the workpiece and the axis. This part only needs to be clamped once to complete the machining.

2）Determine the machining route

（1）Flatten the end face.

（2）Machining can be completed by using G73 loop command according to the workpiece profile.

3）Fill in the machining tool card and process card（see Table 2-2-1）

Table 2-2-1 Machining Tool Card and Process Card

Part drawing No.	2-2-1	CNC lathe process card	Machine model	CKA6150
Part name	Arc and taper Part		Machine tool number	
Table of turning tools			Table of measuring tools	
Tool number	Tool compensation number	Name of turning tools	Turning tool parameters	Measuring tool name	Specification/mm
T01	01	93° cylindrical turning tool	Type D blade	Vernier caliper Micrometer	0-150/0.02 25-50/0.01
Working procedure	Process content		Cutting amount			Processing property
		$S/$（r/min）	$F/$（mm/r）	a_p/mm	
CNC lathe	Cylindrical turning and radial turning for determining reference	500	—	1	Manual
1	Cylindrical turning and radial turning for determining reference	500	—	1	Manual
2	Machining R12 mm arc	600-1 000	0.1-0.2	0.5-3	Automatic
3	Machining ϕ30 mm cylindrical	600-1 000	0.1-0.2	0.5-3	Automatic
4	Machining taper	600-1 000	0.1-0.2	0.5-3	Automatic
5	Machining ϕ36 mm cylindrical	600-1 000	0.1-0.2	0.5-3	Automatic
6	Machining R25 mm arc	600-1 000	0.1-0.2	0.5-3	Automatic
7	Machining R50 mm arc	600-1 000	0.1-0.2	0.5-3	Automatic

Continue Table

8	Machining $\phi 34$ mm cylindrical	600-1 000	0.1-0.2	0.5-3	Automatic
9	Machining taper	600-1 000	0.1-0.2	0.5-3	Automatic

III. Program preparation and machining

1. Taper calculation

The taper is one of the common forms of turning. The common parameters are, as shown in Fig. 2-2-2, the diameter D of the bigger taper end, the diameter d of the smaller taper end, the length L of the taper and the taper ratio C. The relationship between them is:

$$C= (D-d) /L$$

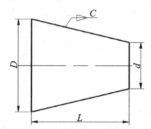

Fig. 2-2-2　Common Parameters for Taper Calculation

2. Rough turning compound loop command (G71)

When the G71 command is used, it is only necessary to specify the finishing route in the program. Each cutting amount command for rough machining will automatically repeat the cutting cooperating with the G70 finishing loop until the part machining is completed. Compared with G01 and G00, the G71 command makes the programming simpler and the program content greatly shortened. It is suitable for turning parts with round bar blanks.

1) Command format

G71 U (Δd) R (e);

G71 P (ns) Q (nf) U (Δu) W (Δw) F (f) S (s) T (t);

Where:

Δd—each cutting depth in X direction (radius value);

e—retract amount;

ns—first segment number of the finishing shape program;

nf—last segment number of the finishing shape program (segment with point B as end point);

Δu—finishing allowance in X direction (diameter value);

Δw—finishing allowance in Z direction.

f, *s*, *t* Any F, S or T function included in the *ns* to *nf* program segments is ignored in the loop, while it is valid in the G71 program segment.

2）G71 command turning route

The parameters in the G71 command segment are shown in Fig. 2-2-3. The NC device first calculates the number of roughly machined tools and the coordinates of each tool route according to the finishing machining route and the depth of each cutting prepared by the user, after reserving the finishing machining allowance in *X* and *Z* directions, the tools cut the allowance in the form of machining outer cylindrical surfaces in layers, and then form a profile similar to the finishing profile. After rough machining, finishing machining can be done using the G70 command.

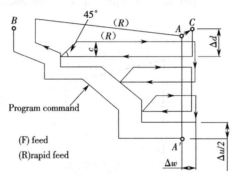

Fig. 2-2-3   Schematic Diagram of G71 Command Line and Parameters

If the finishing shape from *A* to *B* in the above figure is determined by the program, then turning the specified area with $\triangle d$（layered cutting depth）, and leaving finishing allowances $\triangle u/2$ and $\triangle w$.

The starting point of the tool is A. This command can realize the rough machining loop with back cutting depth as $\triangle d$ and finishing allowances as $\triangle u/2$ and $\triangle w$. Among them, $\triangle d$ represents back cutting depth（radius value）, which has no positive and negative signs, and the cutting direction of the tool depends on the *AA'* direction; *e* represents retract amount, which can be set by parameters; *ns* specifies the sequence number of the first program segment of the finishing route, and *nf* specifies the sequence number of the last program segment.

3. Finishing loop（G70）

1）Command format

G70 P（*ns*）Q（*nf*）;

Where:

ns—segment number of the first segment of the finishing program.

nf—segment number of the last segment of the finishing program.

For example: G70 P10 Q20;

2）Turning trajectory of G70 command

When G70 commands the workpiece to be turned, the tool will cut along the actual profile of the workpiece, and the tool will automatically return to the starting point after the loop is completed.

Attention：

The G70 command is the finishing turning loop command, so it is necessary to cooperate with the rough turning loop commands（G71, G72, G73）when using, and can not be used separately.

The F and S values during G70 execution are specified by F and S between segment numbers "*ns*" and "*nf*".

3）Programming example

The part is shown in Fig. 2-2-4, and the blank material is $\phi45$mm$\times78$ mm.

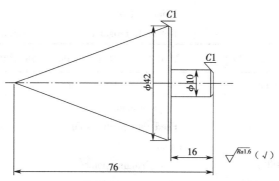

Fig. 2-2-4　Plumb Bob

According to the part shown in Fig. 2-2-4, the machining route of the workpiece is analyzed, and the clamping scheme during machining, together with the tool and cutting amount, is determined. It is divided into two parts according to the machining content, and two corresponding programs are prepared to complete the machining. Table 2-2-2 shows the program for machining the right end of the plumb bob. Table 2-2-3 shows the program for machining the left end of the plumb bob.

Table 2-2-2　Program for Machining the Right End of the Plumb Bob

Program content	Program description
02902;	Program name;
N010 G99 M03 S600 T0101;	Feed rate mm/r, spindle forward rotation 600 r/min, select #1 tool;
N020 G00 X100. Z100.;	Tool changing point;
N030 G00 X50. Z2.;	Starting point of loop;
N040 G71 U2. R0.5;	Give cutting depth 2 mm and retract amount 0.5 mm;

Continue Table

Program content	Program description
N050 G71 P60 Q70 U0.5 W0 F0.2;	Finishing turning route is N060-N070;
N060 G00 X0;	
G01 Z0 F0.1;	
X8.;	
X10. Z−1.;	
Z−16.;	
X41.37;	
X42.37 Z−16.5;	
N070 X45.;	
N090 G70 P60 Q70;	Retract tool;
N100 G00 X100. Z100.;	Finishing turning;
N110 M05;	Return to tool changing point;
N120 M30;	Spindle stop;
	Return to program start after program end

Table 2-2-3　Program for Machining the Left End of the Plumb Bob

Program content	Program description
02903;	Program name;
N010 G99 M03 S600 T0101;	Feed rate mm/r, spindle forward rotation 600 r/min, select #1 tool;
N020 G00 X100. Z100.;	Tool changing point;
N030 G00 X50. Z2.;	Starting point of loop;
N040 G71 U2. R0.5;	Give cutting depth 2 mm and retract amount 0.5 mm;
N050 G71 P60 Q70 U0.5 W0 F0.2;	Finishing turning route is N060-N070;
N060 G00 X0;	
G01 Z0 F0.1;	
X42.37 Z60.5;	
N070 X45.;	Retract tool;
N080 G70 P60 Q70;	Finishing turning;
N090 G00 X100. Z100.;	Return to tool changing point;
N100 M05;	Spindle stop;
N110 M30;	Return to program start after program end

4. Pattern repeating rough turning loop G73

It is mainly used to cut the profile of the fixed trajectory. This compound loop can efficiently cut the workpiece formed by forging, casting or rough turning. For workpieces without similar forming conditions, using this command will instead increase the idle travel of the tool during cutting and reduce efficiency.

1）The command format

G73 U（Δi）W（Δk）R（d）;

G73 P（ns）Q（nf）U（Δu）W（Δw）F（f）S（s）T（t）;

Note:

Δi—total allowance for radial removal during rough cutting（radius value）;

Δk—total allowance for axial removal during rough cutting;

d—number of cycles.

Other parameters have the same meaning as G71.

2）Turning trajectory of G73 command

Its feed path is shown in Fig. 2-2-5. When performing the G73 function, the trajectory shapes of each cutting route are the same, but the positions are different. After each feed, the cutting trajectory is moved to a position of the workpiece. Therefore, it can process the blanks that have been preliminarily formed by rough machining such as forging and casting with high efficiency.

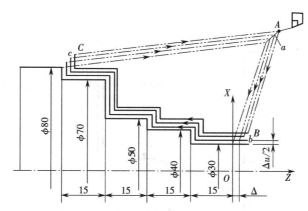

Fig. 2-2-5　Feed Path for G73 Command

3）Examples of programming are shown in Fig. 2-2-6 and the programs are shown in Table 2-2-5. Attention shall be paid to the following issues when machining part in Fig. 2-2-6

（1）Tool selection When machining the part, first pay attention to the tool minor cutting edge angle which should be large so as to avoid overcutting in the machining process such as sharp tool.

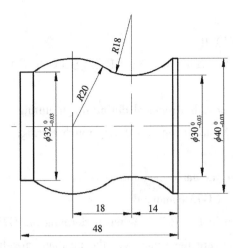

Fig. 2-2-6   Machining of Concave-convex Arc Parts

（2）For the convenience of basic point calculation in this case，the basic points are captured with CAD，as shown in Fig. 2-2-7.

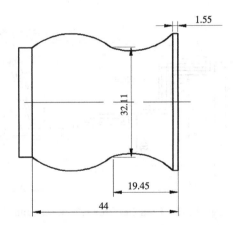

Fig. 2-2-7   Capturing Basic Points with CAD

Table 2-2-4   Processing Program of Concave-convex Arc Parts

Program content	Program description
02111；	Program number；
N010 G99 M03 S600 T0101 ；	Feed rate mm/r，spindle forward rotation 600 r/min，select #1 tool；
N020 G00 X100. Z100.；	Tool changing point；
N030 G00 X47. Z5.	Starting point of loop；
N040 G73 U7.5 R5；	
N050 G73 P60 Q70 U1.0	Cutting depth 2 mm，tool retraction 0.5 mm；
	Fine turning N60 to N070，fine turning allowance 1 mm in X-direction and 0.1 mm in Z-direction；

Continue Table

Program content	Program description
N060 G00 X0;	Approaching workpiece;
G01 Z0 F0.1;	Starting point of end face flattening;
X40.0;	Flatten end face;
Z-1.55;	Machining ϕ40 mm outer circle;
G02 X32.11 Z-19.45 R18.;	Machining R18 mm arc;
G03 X32. Z-44. R20.;	Machining R20 mm arc;
G01 Z-48.;	Machining ϕ32 mm outer circle;
N070 X47.;	Retract tool;
N080 G70 P60 Q70;	Finishing command;
N090 X100.0 Z100.0;	Return to tool changing point;
N100 M05;	Spindle stop;
N110 M30;	End of main program and return

5. Program for Fig. 2-2-1

According to the parts shown in Fig. 2-2-1, the machining route of the workpiece is analyzed, and the clamping scheme during machining, together with the tool and cutting amount, is determined, and the relevant part is formed according to the process content.

Table 2-2-5 shows the program for machining arc and taper parts.

Table 2-2-5　Program for Machining Arc and Taper Parts

Program content	Program description
02112;	Program number;
N010 G99 M03 S600 T0101;	Feed rate mm/r, spindle forward rotation 600 r/min, select #1 tool;
N020 G00 X100. Z100.;	Tool changing point;
N030 G00 X62.0 Z5.0	Starting point of loop;
N040 G73 U21. R11;	Cutting depth 4 mm;
N050 G73 P060 Q070 U1.0 W0.1 F0.2;	
N060 G00 X17.89 Z2.;	Fine turning N060 to N070, fine turning allowance 1
G01 Z0 F0.1;	mm in X-direction and 0.1 mm in Y-direction;
G03 X24. Z-8. R12;	Starting point of machining profile;
G01 X30.;	
Z-25.;	
X36. Z-35.;	
Z-45.;	
G02 X32.83 Z-55.41 R25.;	
G03 X39.12 Z-89.84 R25.;	
G02 X34. Z-108 R15.;	
G01 Z-113;	

Continue Table

Program content	Program description
X48.　Z–133.; 　　　Z–150.; N070 G00 X62.0;	
N080 G70 P060 Q070;	Finishing command;
N090 X100.0　Z100.0;	Return to tool changing point;
N100 M05;	Spindle stop;
N110 M30;	End of main program and return

6. Parts machining

（1）The teacher demonstrates the whole preparation and machining process of the workpiece.

（2）Explain and demonstrate the guarantee method of dimensional tolerance

（3）The students complete the machining of the workpiece in groups, and the teacher gives tour guidance to find and correct the problems of students arising from machining in time.

IV. Task assessment

Task evaluation as per Table 2-2-6.

Table 2-2-6　Scoring Criteria for Arc and Taper Parts

Assessment item		Assessment requirement	Assigned score	Scoring criteria	Test result	Score	Remark
1	Outer circle	$\phi 30_{-0.03}^{0}$ mm	7	Deduct 5 points for each 0.01 mm out of tolerance			
2		$\phi 36_{-0.03}^{0}$ mm	7	Deduct 5 points for each 0.01 mm out of tolerance			
3		$\phi 48_{-0.03}^{0}$ mm	7	Deduct 5 points for each 0.01 mm out of tolerance			
4		$\phi 34$ mm	7	Deduct 5 points for each 0.01 mm out of tolerance			

Continue Table

Assessment item		Assessment requirement	Assigned score	Scoring criteria	Test result	Score	Remark
5	Length	8 mm，17 mm，10 mm（two places），9 mm，30 mm，24 mm，5 mm，17 mm，150 mm	2×10	Full deduction for out-of-tolerance			
6	Arc	R12 mm	5	Full deduction for out-of-tolerance			
7		R25 mm	5	Full deduction for out-of-tolerance			
8		R15 mm	5	Full deduction for out-of-tolerance			
9		φ50 mm	5	2 points will be deducted for one level degraded			
10	Taper	Two places	4×2	Full deduction for out-of-tolerance			
11	End face	Two places	2×2	Full deduction for out-of-tolerance			
12	Surface	Ra1.6 μm （three places）	2×3	2 points will be deducted for one level degraded			
		Ra1.6 μm （seven places）	2×7	2 points will be deducted for one level degraded			
13	Process，program	Provisions on process and program		Deduct 1-5 points for violation of regulations			
14	Standard operation	Rules related to standard operation of CNC lathe.		Deduct 1-5 points for violation of regulations			
15	Safe and civilized production	Rules related to safe and civilized production.		Deduct 1-5 points for violation of regulations			

Assessment item	Assessment requirement	Assigned score	Scoring criteria	Test result	Score	Remark
Summary and feedback during students task implementation:						
Teacher's review:						

Task III　Programming and machining of triangular thread part

I. Drawings and technical requirements

A thread is a common structure on parts, and a part with threads is one of the important parts in machinery and equipment. As a standard part, it is widely used for connection, transmission, and fastening. Threads can be divided into connecting threads and transmission threads by purpose. Thread machining on the numerically controlled lathe of the FANUC system can be programmed with G32 command and G92 command, but each programming method has its own characteristics. Taking Fig. 2-3-1 Stud as an example, this task focuses on the programming method related to threads.

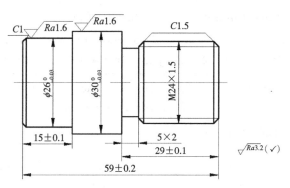

Fig. 2-3-1　Stud

II. Analysis of drawings

1. Analysis of drawings and process arrangement

Fig. 2-3-1 shows a stud, which is typical of threaded parts. The main machining items are external threads of outer circles, grooves, and triangular cylinders. The main knowledge to be mastered during machining is about thread machining and relevant programming commands.

2. Process arrangement

1）Determine the clamping scheme of the workpiece

According to the shape of the blank, the scroll chuck can be selected for clamping, and the program origin is set at the intersection of the right end face of the workpiece and the axis.

Machining of this part can only be completed after two times of clamping. For the first time, the right end of the workpiece is clamped to turn a $\phi30$ mm outer circle and a $\phi26$ mm outer circle. For the second time, take the $\phi26$ mm finely turned outer circle as the positioning reference, flatten the end face, determine the total length, turn a $\phi24$ mm outer circle and a 5×2 tool retraction groove, and then machine the outer threads of the M24×1.5 triangular cylinder.

2）Determine the machining route

（1）Flatten the end face

（2）Roughly and finely turn the $\phi30$mm outer circle and $\phi26$mm outer circle

（3）Turn the workpiece around and clamp the $\phi26$mm outer circle

（4）Flatten the end face（determine total length）

（5）Roughly and finely turn the $\phi24$mm outer circle and 5 mm×2 mm tool retraction groove.

（6）Roughly and finely turn the external threads of the M24×1.5 triangular cylinder

3）Fill in the Machining Tool Card and Process Card（see Table 2-3-1）

Table 2-3-1 Machining Tool Card and Process Card

Part drawing No.	2-3-1	CNC lathe process card		Machine model	CKA6150
Part name	Stud	Machine tool number			
Table of turning tools				Table of measuring tools	
Tool number	Tool compensation number	Name of turning tools	Turning tool parameters	Measuring tool name	Specification/ mm
T01	01	93° cylindrical turning tool	Type D blade	Vernier caliper Micrometer	0-150/0.02 25-50/0.01
T02	02	Grooving tool	Width of tool 4 mm	Vernier caliper Micrometer	0-150/0.02 25-50/0.01
T03	03	Threading tool	60° triangular external threading tool	Vernier caliper	0-150/0.02

Working procedure	Process content	Cutting amount			Processing property
		S/（r/min）	F/（mm/r）	a_p/mm	
CNC lathe	Cylindrical turning and radial turning for determining reference	500	—	1	Manual
1	Machining ϕ30 mm cylindrical	600-1 000	0.1-0.2	0.5-3	Automatic
2	Machining ϕ26 mm outer circle	600-1 000	0.1-0.2	0.5-3	Automatic
CNC lathe	Turning around and clamping ϕ26 mm outer circle, flattening end face, ensuring total length	500	—	1	Manual
1	Machining ϕ24 mm outer circle	600-1 000	0.1-0.2	0.5-3	Automatic
2	Machining 5 mm×2 mm tool retraction groove	600-1 000	0.1-0.2	0.5-3	Automatic
3	Machining external threads of M24×1.5 triangular cylinder	600-800	1.5	—	Automatic

III. Program preparation and machining

1. Clamping scheme for groove workpieces

According to the width of the groove and other conditions, the direct forming method is often adopted during groove cutting, i.e. the width of the groove is the width of the grooving tool edge, i.e. the back cutting depth a_p. This method produces a large cutting force during cutting. In addition, most grooves are located on the outer surface of the part, and the direction of the main cutting edge is parallel to the axis of the workpiece during grooving, which will affect the clamping stability of the workpiece. Therefore, the following two clamping methods can be generally adopted for groove machining on the NC lathe.

(1) Use soft jaws and increase the length of the clamping surface appropriately to ensure accurate positioning and stable clamping.

(2) Use the tailstock and center as auxiliary support and adopt the method of clamping one position and pressing against one position to ensure the clamping stability of the part to the greatest extent.

2. Tool selection and grooving method

1) Selection of grooving tool

Generally, high-speed steel grooving tools and cutting tools with indexable inserts are often used. See Fig. 2-3-2 for the geometry and angle of the grooving tool. The selection of groove cutter is mainly based on two aspects: first, the width a of the grooving tool should be appropriate; second, the length of the cutting edge L should be greater than the depth of the groove. Fig. 2-3-3 shows an external groove cutting tool with indexable inserts.

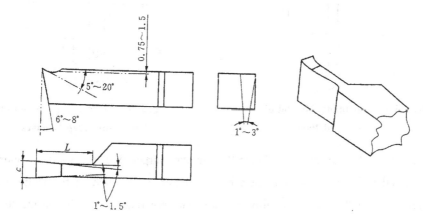

Fig. 2-3-2   High-speed steel grooving tool

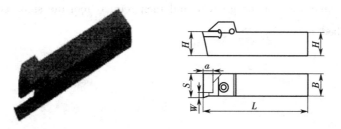

Fig. 2-3-3   Cutting Tool with Indexable Inserts

2）Methods of grooving

（1）For grooves with small width, depth and low accuracy requirements, the method of one-time direct forming with a tool equal to the width of the groove can be used, as shown in Fig. 2-3-4. After the tool cuts into the bottom of the groove, the tool can be paused for a short time using a delay command to trim the roundness of the groove bottom, and the feeding speed can be used during retraction.

（2）For parts with deep grooves with small width but large depth, in order to avoid squeezing and breakage of the turning tool caused by excessive pressure at the front of the tool due to poor chip removal during grooving, the tool shall be driven in several times, and the tool shall stop cutting and return for a distance after cutting into a certain depth of the workpiece, so as to achieve the purpose of breaking and removing cuttings, as shown in Fig. 2-3-5. At the same time, the tool with high strength shall be selected whenever possible.

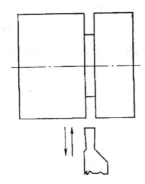

Fig. 2-3-4　Machining Method of
Simple Grooved Parts

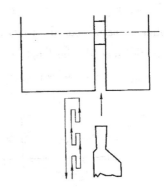

Fig. 2-3-5　Machining Method of
Parts with Deep Grooves

（3）Cutting of wide grooves. Generally, grooves larger than the width of one tool are generally called wide grooves, and the accuracy requirements such as the width and depth of wide grooves and the surface quality requirements are relatively high. During the cutting of a wide groove, rough cutting is often carried out with gang tools, and then a fine grooving tool is used to cut to the groove bottom along one side of the groove and finely machine the bottom and the other side of the groove, and then retract from the side, see Fig. 2-3-6 for the cutting method.

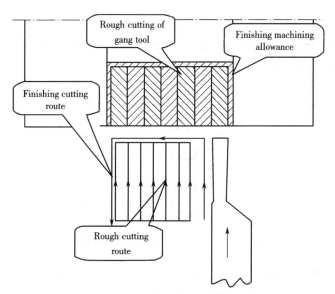

Fig. 2-3-6 Cutting Method of Wide Groove

3. Selection of cutting amount and cutting fluid

The back cutting depth, feed rate, and cutting speed are three factors of the cutting amount. During grooving, the back cutting depth is affected by the tool width, and its adjustment range is small. To increase cutting stability and improve cutting efficiency, the appropriate cutting speed and feed rate shall be selected. For grooving on ordinary lathes, the cutting speed and feed rate shall be lower than those of the outer circle cutting, generally 30% to 70% of those parameters of the outer circle cutting. The performance indexes of the NC lathe are much higher than those of the ordinary lathe. For the selection of the cutting amount, a relatively high speed can be selected. The cutting speed can be selected as 60% to 80% of the cutting speed of the outer circle and the feed rate can be selected from 0.05 to 0.3 mm/r.

It should be noted that vibration is likely to occur during grooving, and this is often caused by an extremely low feed rate or improper matching between the linear rate and the feed rate, and needs to be adjusted in time to ensure the stability of cutting.

During grooving, in order to solve the problem of a small area of the grooving tool bit, which causes poor heat dissipation and is likely to produce high temperature and reduce the cutting performance of the blade, the emulsion type cutting fluid with good cooling performance can be injected to fully cool the tool.

4. Programming command for grooving (cutting off)

For general merely cutting of straight grooves or cutting-off, G01 command can be used, and for wide groove or multi-groove machining, the subprogram and compound loop command can be used for programming.

Example of G0l grooving programming.

As shown in Fig. 2-3-7，cut straight grooves 5 mm wide and machine two chamfers 0.5 mm wide. The grooving tool is 4 mm wide.

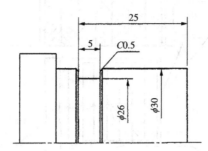

Fig. 2-3-7   Grooving under G01 Command

（1）Programming route and process：The origin of the workpiece is set on the right end face，and the tool alignment point of the grooving tool is the left tool position. As the grooving tool width is less than the groove width and chamfering is required with the grooving tool；thus，machining of the groove needs to be done in three times. The processing route is shown in Fig. 2-3-8.

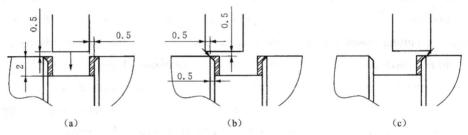

Fig. 2-3-8   G01 Grooving Steps

（2）Program. As shown in Fig. 2-3-8（a），cut from the middle to the bottom of the groove and turn around to retract. The coordinates of the left tool point in Z-direction shall be −24.5 mm.

N010 T0202 M03 S500；

N020 G00 X31. Z−24.5；

N030 G01 X26. F0.05；

N040 X31.；

As shown in Fig. 2-3-8（b），chamfer the left corner and cut the left allowance of the groove before retraction. The starting point of the tool is set on the chamfer extension line；a 0.5 mm distance shall be added in both X-direction and Z-direction. The left tool point shall move to the left with a side allowance of 0.5 mm + chamfer width of 0.5 mm + starting point extension of 0.5 = 1.5 mm.

N050 W−1.5;

N060 X29. W1;

N070 X26.;

N080 W0.5;

N090 X31.;

As shown in Fig. 2-3-8 (c), chamfer the right corner and cut the allowance on the right side of the groove, and move the tool to the center of the groove to retract. The tool shall move 1.5 mm to the right.

N100 W1.5;

N110 X29. W−1.;

N120 X26.;

N130 W−0.5;

N140 X31.;

N150 G00 X100. Z100.;

N160 M05 M30;

5. Knowledge related to threaded workpieces

When machining threads with a CNC lathe, the CNC system controls the pitch and its accuracy, thus simplifying the calculation, requiring no manual replacement of the change gear; what's more, high pitch accuracy ensures no mis-threading. The turning tool moves rapidly during the thread cutting return, and the cutting efficiency is greatly improved. The special CNC thread cutting tool and high cutting speed selected further improve the shape and surface quality of threads.

In thread machining, the back cutting depth ap is equal to the depth of the threading tool into the surface of the workpiece, and the back cutting depth is increasing gradually with each time of cutting by the threading tool. Affected by the size and depth of the thread profile section, the back cutting depth of threads may be very large. Therefore, the cutting speed and feed rate must be selected reasonably.

Table 2-3-2   Feed Times and Amount of Feed for Common Thread Cutting

Unit: mm

Metric thread							
pitch p/mm	1.0	1.5	2.0	2.5	3.0	3.5	4.0
Depth of tooth h/mm	0.649	0.974	1.299	1.624	1.949	2.273	2.598
Back cutting depth and times of cutting — 1 time	0.7	0.8	0.9	1.0	1.2	1.5	1.5
2 times	0.4	0.6	0.6	0.7	0.7	0.7	0.8
3 times	0.2	0.4	0.6	0.6	0.6	0.6	0.6
4 times		0.16	0.4	0.4	0.4	0.6	0.6
5 times			0.1	0.4	0.4	0.4	0.4
6 times				0.15	0.4	0.4	0.4
7 times					0.2	0.2	0.4
8 times						0.15	0.3
9 times							0.2

Note: The recommended values for back cutting depth and cutting times are given in the table, and the programmer can select the suitable values according to his own experience and the actual conditions.

6. Thread programming command

This command is able to machine straight threads and taper threads cyclically. The application is similar to the G90 outer circle loop command.

1) Command format

G00 X_ Z_ (starting point of loop)

G92 X (U) _ Z (W) _ F_

Where:

X_ Z_—coordinate values of end point of thread cutting;

U, W—coordinate increment of the thread cutting end point relative to the starting point of loop;

F—Lead for thread, pitch for single thread.

2) The movement route is shown in Fig. 2-3-9

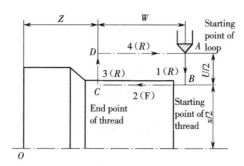

Fig. 2-3-9   Route for Straight Thread Cutting Loop

Going fast from the starting point of loop to the starting point of thread (determined by the starting point Z of loop and the end point X of cutting) —Thread cutting to the end point of the thread—Quick retraction in X-direction—Quick return to starting point of loop in Z-direction.

3) Examples of programming are shown in Fig. 2-3-9 and the program is shown in Table 2-3-3

Table 2-3-3   Cyclic Cutting Program for Straight Thread

Program content	Program description
N3;	Third program segment number (thread machining segment);
N010 G95 M03 S600 T0303	Feed rate setting: mm/r, spindle forward rotation, 600 rpm, #3 tool
N020 G00 X32 Z3	selected;
N030 G92 X29.1 Z–22 F2	Starting point of loop;
N040 X28.5 Z–22	Thread cutting loop 1, feed 0.9 mm;
N050 X27.9 Z–22	Thread cutting loop 2, feed 0.6 mm;
N060 X27.5 Z–22	Thread cutting loop 3, feed 0.6 mm;
N070 X27.4 Z–22	Thread cutting loop 4, feed 0.4 mm;
N080 X100 Z100	Thread cutting loop 5, feed 0.1 mm;
N090 M05	Return to tool changing point;
N100 M30	Spindle stop;
	Return to program start after program end

7. Fig. 2-3-1 Program

According to the parts shown in Fig. 2-3-1, the processing route of the piecework is analyzed, and the clamping scheme, tool, and cutting amount during machining are determined. According to the process content, different parts are formed, and each corresponding program is compiled to complete machining.

Table 2-3-4 shows the program for machining the left end of the stud.

Table 2-3-5 shows the program for machining the right end of the stud.

Table 2-3-4 Program for Machining the Left End of Stud

Program content	Program description
01111;	Program number;
N010 G99 M03 S600 T0101;	Feed rate mm/r, spindle forward rotation 600 r/min, select #1 tool;
N020 G00 X100.Z100.;	Tool changing point;
N030 G00 X30.5 Z2.;	Workpiece approach point;
N040 G01 X30.5 Z−30. F0.2;	Rough turning $\phi30$ mm cylindrical, leave 0.5 mm finishing allowance;
N050 G01 X32. Z−30.;	X-direction tool retraction;
N060 G00 X32.Z2.;	Z-direction tool retraction;
N070 G01 X28.5 Z2. F0.2;	Starting point of first tool in rough turning of $\phi26$ mm outer circle;
N080 G01 X28.5 Z−30.;	End point of first tool in rough turning of $\phi26$ mm outer circle;
N090 G01 X32. Z−30.;	X-direction tool retraction;
N100 G00 X32. Z2.;	Z-direction tool retraction;
N110 G00 X26.5 Z2.;	Starting point of second tool in rough turning of $\phi26$ mm outer circle;
N120 G01 X26.5 Z−30. F0.2;	End point of second tool in rough turning of $\phi26$ mm outer circle;
N130 G01 X32. Z−30.;	X-direction tool retraction;
N140 G00 X32. Z2. S1000;	X-direction tool retraction, finish machining speed 1 000 r/min;
N150 G01 X24. Z2.;	Workpiece approach point;
N160 G01 X24. Z0. F0.1;	Starting point of chamfer;
N170 G01 X26. Z−1.;	End point of chamfer;
N180 G01 X26. Z−15.;	Fine turning of $\phi26$ mm outer circle;
N190 G01 X30. Z−15.;	Finishing turning $\phi30$ mm cylindrical starting point;
N200 G01 X30. Z−30.;	Finishing turning $\phi30$ mm cylindrical end point;
N210 G01 X32. Z−30.;	X-direction tool retraction;
N220 G00 X100. Z100.;	Return to tool changing point;
N230 M05;	Spindle stop;
N240 M30;	Return to program start after program end

Table 2-3-5 Program for Machining Right End of Stud

Program content	Program description
02222;	Program number;
N010 G99 M03 S600 T0101;	Feed rate mm/r, spindle forward rotation 600 r/min, select #1 tool;
N020 G00 X100.Z100.;	Tool changing point;
N030 G00 X32. Z2.;	Workpiece approach point;
N040 G01 X30. Z−29. F0.2;	First tool in rough turning of $\phi24$ mm outer circle;
N050 G01 X32. Z−29.;	X-direction tool retraction;
N060 G00 X32.Z2.;	Z-direction tool retraction;
N070 G00 X28. Z2.;	Starting point of second tool in rough turning of $\Phi24$ outer circle;

Continue Table

Program content	Program description
N080 G01 X28. Z−29. F0.2;	End point of second tool in rough turning of $\Phi 24$ outer circle;
N090 G01 X32. Z−29.;	X-direction tool retraction;
N100 G00 X32. Z2.;	Z-direction tool retraction;
N110 G00 X26. Z2.;	Starting point of third tool in rough turning of $\phi 24$ mm outer circle;
N120 G01 X26. Z−29. F0.2;	End point of third tool in rough turning of $\phi 24$ mm outer circle;
N130 G01 X32. Z−29.;	X-direction tool retraction;
N140 G00 X32. Z2.;	Z-direction tool retraction;
N150 G01 X24.5 Z2.;	Starting point of fourth tool in rough turning of $\phi 24$ mm outer circle;
N160 G01 X24.5 Z−29. F0.2;	End point of fourth tool in rough turning of $\phi 24$ mm outer circle;
N170 G01 X32. Z−29.;	X-direction tool retraction;
N180 G01 X32. Z−2.;	Z-direction tool retraction;
N190 G01 X21. Z2.;	Workpiece approach point;
N200 G01 X21. Z0. F0.1;	Starting point of chamfer;
N210 G01 X23.7. Z−1.5;	End point of chamfer;
N220 G01 X23.7. Z−29.;	Starting point of fine turning of $\phi 24$ mm outer circle;
N230 G01 X32. Z−29.;	End point of fine turning of $\phi 24$ mm outer circle;
N240 G00 X100. Z100.;	Return to tool changing point;
N250 T0202;	Change to 5 mm wide grooving tool;
N260 G00 X32. Z2.;	Workpiece approach point;
N270 G00 X32. Z−29.;	Starting point of grooving;
N280 G01 X20. Z−29. F0.1;	End point of grooving;
N290 G04 X2.;	2 s pause;
N300 G01 X21. Z−29.;	Retract tool;
N310 G01 X24. Z−27.5;	Starting point of chamfer;
N320 G00 X32. Z−27.5;	End point of chamfer;
N330 G00 X100. Z100.;	Return to tool changing point;
N340 T0303;	Change to triangular external threading tool;
N350 G00 X26.Z2.;	Starting point of loop;
N360 G92 X22.9 Z−27. F1.5;	First tool in threading;
N370 X22.3;	Second tool in threading;
N380 X22.15;	Third tool in threading;
N390 X22.05;	Fourth tool in threading;
N400 X22.05;	Fine turning of threads;
N410 G00 X100. Z100.;	Return to tool changing point;
N420 M05;	Spindle stop;
N430 M30;	Return to program start after program end

8. Parts machining

（1）The teacher demonstrates the whole preparation and machining process of the workpiece.

（2）Explain and demonstrate the guarantee method of dimensional tolerance

（3）The students complete the machining of the workpiece in groups，and the teacher gives tour guidance to find and correct the problems of students arising from machining in time.

（4）Operational precautions.

① In order to ensure the consistency of the machining reference，during setting of multiple tools，one reference can be machined with one tool first，and the other tools are respectively set based on the reference.

② The spindle speed and "multiplying power" shall not be changed when machining the threads；otherwise，mis-threading will be caused.

IV. Task assessment

Task evaluation as per Table 2-3-6.

Table 2-3-6　Scoring Criteria for Stud

Assessment item		Assessment requirement	Assigned score	Scoring criteria	Test result	Score	Remark
1	Outer circle	$\phi 30_{-0.03}^{0}$ mm	10	Deduct 5 points for each 0.01 mm out of tolerance			
2		$\phi 30_{-0.03}^{0}$ mm	10	Deduct 5 points for each 0.01 mm out of tolerance			
3	Length	15 mm, 29 mm, 59 mm	5×3	Full deduction for out-of-tolerance			
4	Thread	Large diameter	5	Full deduction for out-of-tolerance			
5		Medium diameter	10	Full deduction for out-of-tolerance			
6		Ra on both sides	10	No score for out-of-tolerance and degradation			
7		Thread angle	10	Template inspection, Full deduction for out-of-tolerance			

Continue Table

Assessment item		Assessment requirement	Assigned score	Scoring criteria	Test result	Score	Remark
8	Thread	*Ra* on both sides	10	Full deduction for out-of-tolerance			
9		C1.5 (two places)	2×2	Full deduction for out-of-tolerance			
10	Chamfer	C1	4	Full deduction for out-of-tolerance			
11	Surface	*Ra*1.6 μm (two places)	6×2	2 points will be deducted for one level degraded			
12	Process, program	Provisions on process and program		Deduct 1-5 points for violation of regulations			
13	Standard operation	Rules related to standard operation of CNC lathe.		Deduct 1-5 points for violation of regulations			
14	Safe and civilized production	Rules related to safe and civilized production.		Deduct 1-5 points for violation of regulations			

Summary and feedback during students task implementation:

Teacher's review:

Item 3

Programming and Machining of
Sleeve Workpiece

[Item description]

The purpose of learning the content of this module is to master the installation method and tool installation method of sleeve parts, to be able to fill in the machining tool card and process card, to be able to complete programming and machining of sleeve parts, as well as to master the inspection method of sleeve parts, and to integrate theory with practice: Carry out simulation machining using CNC simulation software in the programming classroom, carry out on-site machining using the machine tool at the training workshop, and complete the machining tasks individually or in groups.

[Learning objectives]

(1) Be able to analyze the CNC turning process of simple sleeve parts.

(2) Master the installation, use, and setting methods of boring tools.

(3) Master the programming method for inner hole machining and be able to complete the machining of simple sleeve parts.

(4) Be able to compile programs and process cards for simple sleeve parts.

Task I   Programming and machining of internal thread part

I. Drawings and technical requirements

The internal threads are mainly matched with the external threads for connection and power transmission. Common internal threads include coarse triangular threads, fine triangular threads, trapezoidal threads, and internal taper threads.

The machining of internal threads is similar to that of external threads. For fine triangular threads, the machining can be carried out in accordance with G32 and G92 commands.

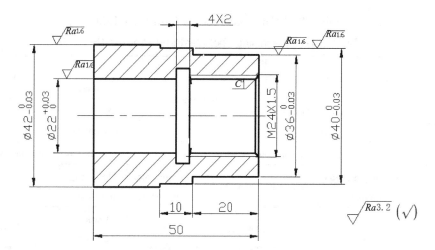

Fig. 3-1-1 Internally Threaded Parts

II. Analysis of drawings and process arrangement

1. Analysis of drawings

As shown in Fig. 3-1-1, there is an internally threaded part, the length of the workpiece is 50 mm, and the dimensions of the three steps of the outer circle are ϕ42 mm, ϕ36 mm, and ϕ40 mm respectively. The size of the inner hole step is ϕ22 mm, the inner thread is M24×1.5, the size of the tool retraction groove is 4 mm×2mm, and the technical requirements are: sharp angle smoothing C1.

2. Process arrangement

1）Determine the clamping scheme for the workpiece

Machining of this part can only be completed with two times of clamping. For the first time, clamp the left end and turn the right end to complete machining of the through hole drilling, ϕ36 mm and ϕ40 mm outer circles, ϕ22 mm inner hole, 4 mm×2 mm inner groove, and M24×1.5 inner thread, and for the second time, machine the ϕ42 mm outer circle with the ϕ36 mm finely turned outer circle as the positioning reference.

2）Determine the machining route

（1）Flatten the end face and drill blank hole ϕ20 mm.

（2）Roughly and finely turn ϕ36 mm and ϕ40 mm outer circles.

（3）Roughly and finely turn ϕ22 mm and ϕ22.5 mm inner holes.

（4）Machine 4 mm×2 mm inner groove.

（5）Machine M24×1.5 internal thread.

（6）Turn the workpiece around and clamp ϕ36 mm outer circle.

（7）Roughly and finely turn ϕ42 mm outer circle.

3）Fill in the Machining Tool Card and Process Card（see Table 3-1-1）

Table 3-1-1 Machining Tool Card and Process Card

Part drawing No.	3-1-1	CNC lathe process card		Machine model	CKA6150
Part name	Internally threaded part			Machine tool number	
Table of turning tools				Table of measuring tools	
Tool number	Tool compensation number	Name of turning tools	Turning tool parameters	Measuring tool name	Specification/mm
T01	01	93° outer circle finishing tool	Type D blade	Vernier caliper Micrometer	0-150/0.01 25-50/0.01
T02	02	Boring tool	T-blade	Bore dial indicator	18-35/0.01
T03	03	Internal grooving tool	4 mm wide blade		
T04	04	Internal threading tool （Fig. 3-1-2）		Feeler gauge	M24×1.5
		Bit ϕ20		Vernier caliper	0-150/0.02

Working procedure	Process content	Cutting amount			Processing property
		$S/$（r/min）	$F/$（mm/r）	$a_p/$mm	
CNC lathe	Turning the end face to determine the reference	500	—	1	Manual
1	Drilling	300			Manual
2	Machining ϕ36 mm and ϕ40 mm outer circles	800-1 000	0.1-0.2	0.5-3	Automatic
3	Machining ϕ22 mm and ϕ22.5 mm inner holes	600-800	0.05-0.1	0.3-1	Automatic
4	Machining 4×2 inner groove	300	0.1	4	Automatic
5	Machining M24×1.5 internal thread	300	1.5	0.05-0.4	Automatic
CNC lathe	Turning around and clamping ϕ36 outer circle，ensuring total length	—	—	—	Manual
1	Machining ϕ42 mm cylindrical	800-1 000	0.1-0.2	0.1-1	Automatic

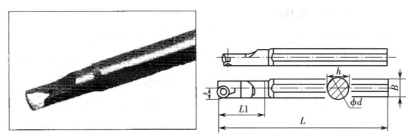

Fig. 3-1-2 Internal Threading Tool T04

III. Program preparation and machining

1. Clamping scheme for sleeve parts

For the inner and outer circles, end faces, and reference axes of the sleeve parts, there are certain shape and position accuracy requirements. The outer circle can be selected as the finishing reference of the sleeve parts, but the center hole and one end face are often used as the finishing reference. For disc parts of different structures, it is impossible to ensure the accuracy requirements of their shape and position with one process scheme.

According to the structural characteristics of sleeve parts, three-jaw chucks, four-jaw chucks, or disc chuck can be used for clamping in machining with a CNC lathe. Due to errors in centering accuracy of three-jaw chucks, it is not suitable for secondary clamping of workpieces with high coaxiality requirements. Three-jaw chucks can be used for clamping sleeve parts that allows machining of the inner and outer circular end faces, chamfering and cutting in one time. Larger parts are often clamped with four-jaw chucks or disc chucks. For finished parts, soft jaws are generally used, and for complex sleeve parts, special fixtures are sometimes used.

2. Tool selection

The tool selection for machining the outer cylindrical surface of the sleeve part is the same as that of the shaft part. The machining of inner holes is one of the characteristics of sleeve parts. According to the process requirements of inner holes, there are many machining methods. The commonly used methods include drilling, bearizing, reaming, boring, grinding, broaching, and lapping. Applicable machining tools should be selected according to different machining methods.

Generally, sleeve parts include the outer circles, tapered surfaces, arcs, grooves, holes, threads, and other structures. According to the machining requirements, the commonly used tools include rough boring tools, fine boring tools, inner groove tools, internal threading tool, center drills, and twist drills.

3. Selection of cutting fluid

Machining of the sleeve parts is more difficult than machining of the shaft parts in CNC machining. Due to the characteristics of the sleeve parts, it is not easy for the cutting fluid to reach the cutting area, the temperature during cutting is high, and the wear of the cutting tool is serious. In order to reduce the machining deformation of the workpiece and improve its machining accuracy, suitable cutting fluid shall be selected according to different workpiece materials, and the filling position of the cutting fluid shall be adjusted in due time.

4. Method of ensuring total length after turning around for machining

In the previous chapters, we have dealt with the machining of half shaft parts, and from this chapter, we are going to talk about machining after turning the workpiece around, which requires us to ensure the total length of the parts. The common method is to flatten both end faces of the blank before machining, and then machine the blank to the required length during re-flattening of the end face. During machining after turning around, allow slight contact between the tool and the end face during the Z-axis tool setting, and then enter the trial cutting value in the tool setting interface, while the end face flattening operation is no longer carried out.

5. Example of Step Hole Workpiece Programming

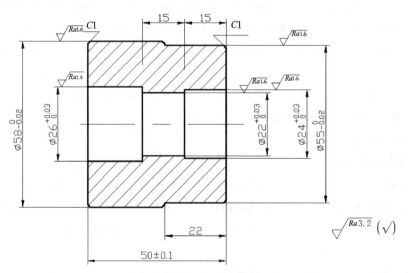

Fig. 3-1-3 Step Hole

Table 3-1-2 Program for Machining Left End of Step Hole

Program content	Program description
O3101;	Program number;
N1;	1st program segment number;
G99 M03 S700 T0202;	2# tool selected, spindle forward rotation, 700 rpm;
G00 X100.0 Z100.0;	Rapid movement to safe point;
G00 X18.0 Z2.0;	Rapid movement to loop point;
M08;	Coolant on;
G71 U1.0 R0.5;	Rough machining ϕ26 mm and ϕ22 mm inner hole
G71 P10 Q20 U−0.5 W0.05 F0.1;	loop;
N10 G00 G41 X27.0;	Machining loop starting section program, tool right
G01 Z0;	compensation;
X26.0 Z−0.5;	
Z−20.0;	
X22.0 C0.5;	
Z−36.0;	Turning length increased by 1 mm;
N20 G00 G40 X18.0;	Machining loop end−point section program, canceling
	tool compensation;
G00 Z100.0;	Rapid movement to safe point;
X100.0;	
M09;	Coolant off;
M00;	Program suspension;
N2;	2nd program segment number;
G99 M03 S800 T0202;	#2 tool selected, spindle forward rotation, 800 rpm;
G00 X100.0 Z100.0;	Rapid movement to safe point;
G00 X18.0 Z2.0;	Rapid movement to loop point;
M08;	Coolant on;
G70 P10 Q20 F0.05;	Finish machining ϕ26 mm and ϕ24 mm inner hole
	loop;
G00 Z100.0;	Rapid movement to safe point;
X100.0;	
M09;	Coolant off;
M30;	Return to program start after program end

Table 3-1-3　Program for Machining Right End of Step Hole

Program content	Program description
03102;	Program number;
N1;	1st program segment number;
G99 M03 S600 T0202;	#2 tool selected，spindle forward rotation，600 rpm;
G00 X100.0 Z100.0;	Rapid movement to safe point;
G00 X18.0 Z2.0;	Rapid movement to loop point;
M08;	Coolant on;
G71 U1.0 R0.5;	Rough machining ϕ24 mm inner hole loop;
G71 P10 Q20 U−0.5 W0.05 F0.1;	Machining loop starting section program，tool right compensation;
N10 G00 G41 X25.0;	
G01 Z0;	
X24.0 Z−0.5;	
Z−15.0;	
X21.0;	
X22.0 Z−15.5;	Machining loop end-point section program，canceling tool
N20 G00 G40 X18.0;	compensation;
G00 Z100.0;	Rapid movement to safe point;
X100.0;	
M09;	Coolant off;
M00;	Program suspension;
N2;	2nd program segment number;
G99 M03 S800 T0202;	#2 tool selected，spindle forward rotation，800 rpm;
G00 X100.0 Z100.0;	Rapid movement to safe point;
G00 X18.0 Z2.0;	Rapid movement to loop point;
M08;	Coolant on;
G70 P10 Q20 F0.05;	Fine machining ϕ30 mm inner hole loop;
G00 Z100.0;	Rapid movement to safe point;
X100.0;	
M09;	Coolant off;
M30;	Return to program start after program end

6. Programming of inner groove workpieces

1）Tools

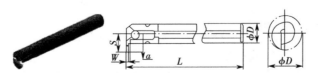

Fig. 3-1-4　Inner Grooving（Groove Cutting）Tool

2) Example of programming

As shown in Fig. 3-1-5, there is a sleeve with inner grooves. The length of the workpiece is 50 mm, and the dimensions of the three steps of the outer circle are $\phi42$ mm, $\phi36$ mm, and $\phi40$ mm respectively. The dimensions of the three steps of the inner hole are $\phi24$ mm, $\phi22$ mm, and $\phi24$ mm respectively, including two inner grooves of 4 mm×2 mm and 10 mm×3 mm. Technical requirement: sharp angle smoothing C0.5.

Table 3-1-4 Program for Machining Right End of Parts with Inner Grooves

Program content	Program description
03301;	Program number;
N1;	1st program segment number;
G99 M03 S600 T0303;	3# tool selected, spindle forward rotation, 600 rpm;
G00 X100.0 Z100.0;	Rapid movement to safe point;
G00 X20.0 Z2.0;	Rapid movement to loop point;
Z-15.0;	
M08;	Coolant on;
G01 X28.0;	Machining 4×2 inner groove;
X20.0;	Retract tool;
G00 Z100.0;	Rapid movement to safe point;
X100.0;	
M09;	Coolant off;
M05;	Spindle stop;
M30;	Return to program start after program end

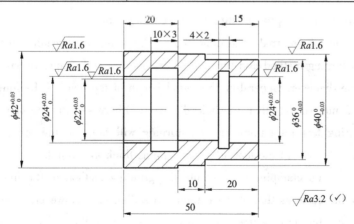

Fig. 3-1-5 Part with Inner Grooves

Table 3-1-5 Program for Machining Left End of Parts with Inner Grooves

Program content	Program description
03302；	Program number；
N1；	1st program segment number；
G99 M03 S600 T0303；	3# tool selected，spindle forward rotation，600 rpm；
G00 X100.0 Z100.0；	Rapid movement to safe point；
G00 X20.0 Z2.0；	Rapid movement to loop point；
Z−20.0；	
M08；	Coolant on；
G01 X30.0；	Machining 10×3 inner groove；
X22.0；	
G00 W4.0；	
G01 X30.0；	
X22.0；	
G00 W2.0；	
G01 X30.0；	
Z−20.0；	
X22.0；	Retract tool；
G00 Z100.0；	Rapid movement to safe point；
X100.0；	
M09；	Coolant off；
M05；	Spindle stop；
M30；	Return to program start after program end

7. Selection and clamping of internal threading tool

（1）Selection of internal threading tool：The internal threading tool is selected according to its turning method and the workpiece material and shape. Its size is limited by the threaded hole diameter. Generally，the radial length of the head of the internal threading tool shall be 3-5 mm smaller than the hole diameter. Otherwise，the top of the threads will be damaged during tool retraction，or even turning will fail. On the premise of ensuring cuttings removal，the size of the tool arbor shall be as thick as possible.

（2）Grinding and clamping of the tool：the grinding method of the internal threading tool is basically the same as that of the external threading tool. However，when grinding the tool nose，note that the halving line must be perpendicular to the tool arbor；otherwise，the tool arbor will collide with and damage the inner hole during turning internal threads. The tool nose width shall meet the requirements，generally 0.1 times the pitch.

8. Calculation of triangular internal thread size

Take the M24×1.5 internal thread as an example.

（1）Thread major diameter D=nominal diameter=24 mm

（2）Threaded hole diameter D_{hole}=nominal diameter D−1.082 5×pitch P=24−

1.082 5×1.5 =22.376 mm

9. Precautions for internal thread machining

（1）Before turning the internal threads, turn the inner holes, end faces, and chamfers of the workpiece properly.

（2）The cutting method of the tool is the same as that for the external threads, and the straight feed method is adopted for threads with a pitch less than 1.5 mm or cast iron threads; the left and right cutting method is adopted for threads with a pitch greater than 2 mm. It is difficult to visually observe the internal threading. Generally, the left and right separate cutting is carried out according to the cuttings removal observed, and the roughness of the thread surface is judged.

10. Program for part machining

According to the parts shown in Fig. 3-1-1, the machining route of the workpiece is analyzed, and the clamping scheme during machining, together with the tool and cutting amount adopted, is determined. According to the process content, five parts are formed, and five programs are compiled correspondingly to complete the machining. Here only the programs for turning the inner hole and machining the inner thread are listed. Table 3-1-6 shows the program for machining the left end of the internally threaded part, and Table 3-1-7 shows the program for machining the right end of the internally threaded part.

Table 3-1-6　Program for Machining Inner Hole of Internally Threaded Parts

Program content	Program description
03009;	Program number;
N1;	1st program segment number;
G99 M03 S600 T0202;	#2 tool selected , spindle forward rotation, 600 rpm;
G00 X100.0 Z100.0;	Rapid movement to safe point;
G00 X19.0 Z2.0;	Rapid movement to loop point;
M08;	Coolant on;
G71 U1.0 R0.5;	Roughing inner hole loop;
G71 P10 Q20 U−0.5 W0.05 F0.1;	
N10 G00 G41 X24.0;	Machining loop starting section program, tool right compensation;
G01 Z0;	
X22.38 Z−1.0;	
Z−25.0;	
X22.0;	
Z−51.0;	

Continue Table

Program content	Program description
N20 G00 G40 X19.0; G00 Z100.0; X100.0;	Machining loop end-point section program, canceling tool compensation; Rapid movement to safe point;
M09;	Coolant off;
M00;	Program suspension;
N2;	2nd program segment number;
G99 M03 S800 T0202;	#2 tool selected, spindle forward rotation, 800 rpm;
G00 X100.0 Z100.0;	Rapid movement to safe point;
G00 X19.0 Z2.0;	Rapid movement to loop point;
M08;	Coolant on;
G70 P10 Q20 F0.05;	Finishing inner hole loop;
G00 Z100.0;	Rapid movement to safe point;
X100.0;	
M09;	Coolant off;
M30;	Return to program start after program end

Table 3-1-7 Program for Machining Right End of Internally Threaded Parts

Program content	Program description
03010;	Program number;
N1;	1st program segment number;
G99 M03 S300 T0404;	#4 tool selected, spindle forward rotation, 300 rpm;
G00 X100.0 Z100.0;	Rapid movement to safe point;
G00 X20.0 Z5.0;	Rapid movement to loop point;
M08;	Coolant on;
G92 X22.8.0 Z-22.5 F1.5;	Start of thread machining loop;
X23.2;	
X23.6;	
X23.9;	
X23.95;	
X24.0;	
X24.0;	
G00 Z100.0;	Rapid movement to safe point;
X100.0;	
M05;	Spindle stop;
M09;	Coolant off;
M30;	Return to program start after program end

11. Parts machining

(1) The teacher demonstrates the whole preparation and machining process of the workpiece.

(2) Explain and demonstrate the guarantee method of dimensional tolerance

(3) The students complete the machining of the workpiece in groups, and the teacher gives tour guidance to find and correct the problems of students arising from machining in time.

(4) Operational precautions

① When machining the internal thread, note that the starting point shall not be too close to the end face of the workpiece to avoid collision between the left tool body and the workpiece.

② When machining the internal thread, gradually reduce the cutting amount time after time, which is similar to that for the external thread, but the cutting amount should not be too large as the internal threading tool body is relatively thin and the extension length is long.

③ Pay attention to the diameter of the tool body when selecting the internal threading tool, and do not allow the tool body to interfere with the workpiece.

12. Parts measurement

There are two types of inspection methods for internal threads: comprehensive inspection and single-item inspection. Generally, a comprehensive inspection is carried out by using a feeler gauge to inspect the comprehensive results of geometric parameter deviations affecting thread interchangeability.

The internal thread feeler gauge is divided into the go end and the no-go end. If the measured internal thread can engage with the go end of the feeler gauge and does not completely engage with the no-go end of the feeler gauge (the thread no-go gauge is only allowed to engage with two sections of the threads under test, and the amount of engagement shall not exceed two pitches), it indicates that the middle diameter of the measured internal thread does not exceed the middle diameter of its maximum material profile, and the single middle diameter does not exceed the middle diameter of its minimum material profile, then the engaging performance and connection strength can be guaranteed, and the middle diameter of the thread under test is accepted; otherwise, it is not accepted.

Fig. 3-1-6   Internal Thread Feeler Gauge

IV. Task assessment

Task evaluation as per Table 3-1-8.

Table 3-1-8   Scoring Criteria for Internally Threaded Parts

Assessment item		Assessment requirement	Assigned score	Scoring criteria	Test result	Score	Remark
1	Outer circle	$\phi42_{-0.03}^{0}$ mm	12	Deduct 6 points for out-of-tolerance 0.01 mm			
		$\phi40_{-0.03}^{0}$ mm	12	Deduct 6 points for out-of-tolerance 0.01 mm			
		$\phi36_{-0.03}^{0}$ mm	12	Deduct 6 points for out-of-tolerance 0.01 mm			
2	Inner hole	$\phi22_{0}^{+0.03}$ mm	15	Deduct 7 points for out-of-tolerance 0.01 mm			
3	Inner groove	4 mm×2 mm	8	No score for nonconformity			
4	Internal thread	M24×1.5	10	No score for nonconformity			
		21 mm	8	No score for nonconformity			
	Length	50±0.1 mm	7	Deduct 4 points for out-of-tolerance 0.1			
	Roughness	Ra3.2 μm （four places）	16	No score for 1 level higher for Ra			
5	Process, program	Provisions on process and program		Deduct 1-5 points for violation of regulations			
6	Specification Operation	Rules related to standard operation of CNC lathe.		Deduct 1-5 points for violation of regulations			

7	Safe and civilized production	Relevant provisions on safe and civilized production	Deduct 1-5 points for violation of regulations		
Summary and feedback during students task implementation:					
Teacher's review:					

Task II   Programming and machining of shaft sleeve part

I. Drawings and technical requirements

As shown in Fig. 3-2-1, there is a bearing sleeve, the length of the workpiece is 60 mm, and the step dimensions of the outer circle are $\phi58$ mm and $\phi45$ mm respectively. The dimension of the inner hole step is $\phi30$ mm, the dimension of the inner groove is 32×20mm; technical requirement: sharp corner smoothing $C1$, $C2$.

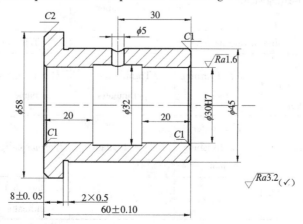

Fig. 3-2-1 Bearing Sleeve Parts

II. Analysis of drawings

1. Analysis of drawings

Fig. 3-2-1 shows a simple bearing part. The surface of the part consists of two steps. The part drawing dimensions are completely marked，meeting the dimension marking requirements of CNC machining. The contour description is clear and complete. The part material is aluminum；it has good machining and cutting performance，while having no requirements for heat treatment or hardness.

2. Process arrangement

1）Determine the clamping scheme of the workpiece

This part is a through-hole sleeve，with high dimensional requirements. The blank above $\phi60$ mm shall be used，with a length of 65 mm. First，machine $\phi58$ mm and $\phi45$ mm outer circle diameters，then machine $\phi30$ mm inner holes and $\phi32$ mm inner grooves. Turn around and align and clamp the $\phi45$ mm outer circle to complete chamfering and the full-length machining.

2）Determine the machining route

（1）Flatten the end face and drill blank hole $\phi28$ mm.

（2）Roughly and finely turn $\phi58$ mm and $\phi45$ mm outer circles.

（3）Roughly and finely turn $\phi30$ mm inner holes.

（4）Machine $\phi32$ mm inner groove.

（5）Turn the workpiece around and align it，and clamp the $\phi45$ mm outer circle.

（6）Machine the full length 60 mm and chamfer the workpiece.

3）Fill in the Machining Tool Card and Process Card（see Table 3-2-1）

Table 3-2-1 Machining Tool Card and Process Card

Part drawing No.	3-2-1	CNC lathe process card Machine tool number	Machine model	CKA6150
Part name	Bearing sleeve part			
Table of turning tools			Table of measuring tools	
Tool number	Tool compensation number	Name of turning tools	Turning tool parameters	Measuring tool name	Specification/ mm
T01	01	93° outer circle finishing tool	Type D blade	Vernier caliper	0-150/0.01
T02	02	Boring tool	T-blade	Bore dial indicator	18-35/0.01

Continue Table

T03	03	Internal grooving tool	4 mm wide blade				
		Bit ϕ28 mm		Vernier caliper		0-150/0.02	
Working procedure	Process content			Cutting amount			Processing property
				$S/$ (r/min)	$F/$ (mm/r)	a_{p}/mm	
CNC lathe	Cylindrical turning and radial turning for determining reference			500	—	1	Manual
1	Drilling			300	—	—	Manual
2	Machining ϕ58 mm and ϕ45 mm outer circles			800-1 000	0.1-0.2	0.5-3	Automatic
3	Machining ϕ30 mm inner hole			600-800	0.05-0.1	0.3-1	Automatic
4	Machining ϕ32 mm inner groove			600	0.1	4	Automatic
CNC lathe	Turning around and clamping ϕ45 mm outer circle			—	—	—	Manual
1	Machining total length 60 mm			800-1 000	0.1-0.2	0.5-3	Automatic

III. Program preparation and machining

1. Rolling bearings construction

Common rolling bearings are generally composed of two rings (i.e. inner ring, outer ring), rolling element, retainer, and other basic elements (Fig. 3-2-2). Generally, the inner ring fits with the journal and rotates with the shaft, and the outer ring is fixed in the bearing seat hole of the retainer. When the inner and outer rings rotate relatively, the rolling element rolls on the raceway of the inner and outer rings; the retainer makes the rolling element evenly distributed and avoids contact friction and wear between adjacent rolling elements.

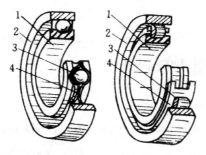

Fig. 3-2-2 Structure of Rolling Bearing

1—Inner ring; 2—Outer ring; 3—Rolling element; 4—Retainer

The inner and outer rings and rolling elements of the rolling bearing are generally made of special rolling bearing steel, such as GCr9, GCr15, and GCr15SiMn. The retainer is usually made of stamped soft material such as low carbon steel plates, or made of copper alloys, plastics, etc.

2. Four basic characteristics of rolling bearing

1) Contact angle

As shown in Fig. 3-2-3, the included angle α between the normal line at the contact of the rolling element and the outer ring in the rolling bearing and the plane perpendicular to the bearing axial lead is called the contact angle. The larger α is, the greater the bearing's ability to withstand axial loads will be.

2) Clearance

The maximum clearance between the rolling element and the inner and outer ring raceways is called the bearing clearance.

3) Angle of displacement

As shown in Fig. 3-2-4, the included sharp angle between the axes of the inner and outer rings of the bearing when the axes are inclined relative to each other is called the angle of displacement. Bearings that can automatically adapt to the angle of displacement are called self-aligning bearings.

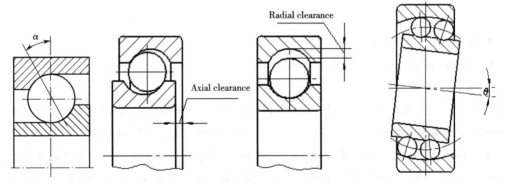

Fig. 3-2-3 Contact Angle and Clearance   Fig. 3-2-4 Angle of Displacement

4) Limit speed

The maximum allowable speed of rolling bearings under certain load and lubrication conditions is called the limit speed, and the specific values are shown in the relevant manuals.

3. Types of rolling bearings

There are many types of rolling bearings, and several common classification methods are described below.

(1) According to the shape of the rolling element, rolling bearings can be divided into two types: ball bearings and roller bearings.

（2）According to the number of rows of the rolling elements, rolling bearings can also be divided into single-row, double-row, and multiple-row rolling bearings.

（3）According to whether self-aligning is available, rolling bearings can be divided into self-aligning bearings and non-self-aligning bearings. The self-aligning bearing allows a large angle of displacement.

（4）Depending on the direction of bearing load, rolling bearings can be divided into two types: radial bearings and thrust bearings.

① Radial bearing: It mainly bears radial loads. The bearing whose nominal contact angle $\alpha=0°$ is called radial contact bearing; The bearing whose nominal contact angle $0°<\alpha_y 45°$ is called angular contact radial bearing. The larger the contact angle, the greater the ability to bear axial loads.

② Thrust bearing: It mainly bears axial loads. The bearing whose nominal contact angle $45°<\alpha<90°$ is called angular contact thrust bearing; The bearing whose nominal contact angle $\alpha=90°$ is called axial contact bearing or thrust bearing. The larger the contact angle, the smaller the ability to bear the radial loads, the larger the ability to bear the axial loads, and the axial thrust bearing can only bear the axial loads.

4. Fig. 3-2-1 Part Program

According to the part shown in Fig. 3-2-1, the machining route of the workpiece is analyzed, and the clamping scheme during machining, together with the tool and cutting amount, is determined. It is divided into two parts according to the machining content, and three corresponding programs are prepared to complete the machining. Here only the programs for the inner hole and the inner groove of the lathe are listed. Table 3-2-2 shows the program for machining the inner hole of bearing sleeve parts and Table 3-2-3 shows the program for machining the inner groove of bearing sleeve parts.

Table 3-2-2 Programs for Machining Inner Hole of Bearing Sleeve Parts

Program content	Program description
03501;	Program number;
N1;	1st program segment number;
G99 M03 S600 T0202;	#2 tool selected , spindle forward rotation, 600 rpm;
G00 X100.0 Z100.0;	Rapid movement to safe point;
G00 X25.0 Z2.0;	Rapid movement to loop point;
M08;	Coolant on;
G71 U1.0 R0.5;	Roughing inner hole loop;
G71 P10 Q20 U-0.5 W0.05 F0.1;	
N10 G00 G41 X28.0;	Machining loop starting section program, tool right compensation;
G01 Z0;	
X30 Z-1.0;	

Program content	Program description
Z–61;	
N20 G00 G40 X25.0;	Machining loop end-point section program, canceling tool compensation;
G00 Z100.0;	Rapid movement to safe point;
X100.0;	
M09;	
M00;	Coolant off;
N2;	Program suspension;
G99 M03 S800 T0202;	2nd program segment number;
G00 X100.0 Z100.0;	#2 tool selected, spindle forward rotation, 800 rpm;
G00 X25.0 Z2.0;	Rapid movement to safe point;
M08;	Rapid movement to loop point;
G70 P10 Q20 F0.05;	Coolant on;
G00 Z100.0;	Finishing inner hole loop;
X100.0;	Rapid movement to safe point;
M09;	Coolant off;
M30;	Return to program start after program end

Table 3-2-3 Programs for Machining Inner Groove of Bearing Sleeve Part

Program content	Program description
03502;	Program number;
N1;	1st program segment number;
G99 M03 S600 T0303;	Select #4 tool, spindle forward rotation, 600 r/min;
G00 X100.0 Z100.0;	Rapid movement to safe point;
X25.0	
G00 X25.0 Z–24.2; (+tool width)	Rapidly move to cycle point (note: plus cutting width 4 mm);
M08;	Coolant on;
G01 X31.8 Z–24.2 F0.05;	Start machining, first cutting;
G00 X25;	
Z–28.2;	Second-time cutting;
G01 X31.8;	
G00 X25;	
Z–32.2;	Third-time cutting;
G01 X31.8;	
G00 X25;	
Z–35.8;	Fourth-time cutting;
G00 X25;	
Z–36;	Finishing turning wide groove;
G01 X32;	

Program content	Program description
Z–24; X25; G00 Z100.0; X100.0;	Rapid movement to safe point;
M05;	Spindle stop;
M09;	Coolant off;
M30;	Return to program start after program end

5. Parts machining

(1) The teacher demonstrates the whole preparation and machining process of the workpiece.

(2) Explain and demonstrate the guarantee method of dimensional tolerance

(3) The students complete the machining of the workpiece in groups, and the teacher gives tour guidance to find and correct the problems of students arising from machining in time.

(4) When machining the inner groove, the cutting force is large, and attention shall be paid to the cuttings discharge and coolant pouring at all times. Any abnormality shall be solved in time.

IV. Part Measurement

1. Internal measuring micrometer

The internal measuring micrometer is shown in Fig. 3-2-5 to measure the small internal diameter and the width of the inner side groove. It is characterized by easy alignment of inner hole diameter and easy measurement. The reading value of the domestic internal measuring micrometer is 0.01mm. There are two common measuring ranges: 5-30 and 25-50 mm. Fig. 3-2-5 shows an internal measuring micrometer of 5-30 mm. The reading method of the internal measuring micrometer is the same as that of the outside micrometer, except that the graduation size on the sleeve is opposite to that of the outside micrometer, and its measurement direction and reading direction are also opposite to that of the outside micrometer.

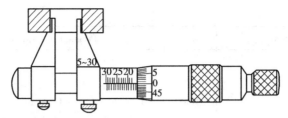

Fig. 3-2-5   Internal Measuring Micrometer

2. Internal micrometer with three-point contact

It is suitable for measuring precision inner holes with medium and small diameters, especially for measuring the diameter of deep holes. Measuring ranges (mm): 6-8, 8-10, 10-12, 11-14, 14-17, 17-20, 20-25, 25-30, 30-35, 35-40, 40-50, 50-60, 60-70, 70-80, 80-90, 90-100. The zero position of the internal micrometer with three-point contact must be calibrated in the standard hole.

The working principle of the internal micrometer with three-point contact: Fig. 3-2-6 shows the micrometer with a measurement range of 11-14 mm. When the ergometric mechanism 6 is rotated clockwise, it will drive the micrometric screw 3 to rotate and move along the spiral direction of the threaded sleeve 4. Therefore, the square taper thread at the end of the micrometer screw will push 1 (three measuring jaws) to move radially. The elastic force of the torsion spring 2 makes the measuring jaws closely attach to the square taper thread and expand and contract with the advance and retreat of the micrometer screw.

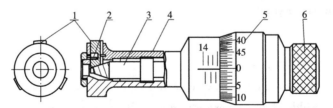

Fig. 3-2-6   Internal Micrometer with Three-point Contact

1—Measuring jaws; 2—Torsion spring; 3—Micrometric Screw;
4—Threaded sleeve; 5—Differential cylinder; 6—Ergometric mechanism

The radial pitch of square taper threads of the internal micrometer with three-point contact is 0.25 mm. That is, when the ergometric mechanism revolves clockwise for one circle, the measuring jaws 1 move outward (in the radius direction) by 0.25 mm, and the circumferential diameter of the three measuring jaws increased by 0.5 mm. That is, when the dial indicator revolves for one circle, the measurement diameter increases by 0.5 mm, and 100 equal divisions are engraved on the circumference of the dial indicator, so its reading value is 0.5 mm ÷ 100=0.005 mm.

3. Application method of bore dial indicator

The bore dial indicator is used to measure the cylindrical hole, which is equipped with a complete set of adjustable measuring heads. Before use, the zero position must be combined and calibrated, as shown in Fig. 3-2-7.

When assembling, install the dial indicator into the connecting rod, so that the short pointer is at the position of 0-1, the long pointer coincides with the axis of the connecting rod, and the words on the dial are vertically downward for observation during measurement and the dial indicator shall be tightened after installation. During rough machining, it is better to first measure with a vernier caliper or an inner caliper. The bore dial indicator is as valuable as other precision measuring instruments, its quality and accuracy directly affect the machining accuracy and service life of the workpiece.

During rough machining, the machined surface of the workpiece is rough and uneven so that the measurement is inaccurate. Therefore, care and maintenance shall be taken, and measurement shall be carried out during finishing.

Before measurement, the dimensions shall be adjusted with an outside dial gauge according to the hole diameter measured, as shown in Fig. 3-2-8. When adjusting the dimension, correctly select the length and extension distance of the replaceable measuring head, and make the measured size in the middle of the overall movement of the movable measuring head.

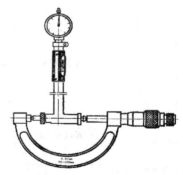

Fig. 3-2-7 Bore Dial Indicator   **Fig. 3-2-8 Dimension Adjustment
                                          with Outside Dial Gauge**

During measurement, the centerline of the connecting rod shall be parallel to the centerline of the workpiece without skewing. At the same time, more points shall be measured on the circumference to find out the actual size of the diameter of the hole and see whether it is within the tolerance range, as shown in Figure 3-2-9.

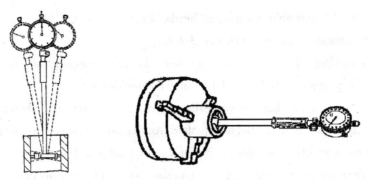

Fig. 3-2-9   Application Method of Bore Dial Indicator

V. Task assessment

Task evaluation as per Table 3-2-4.

Table 3-2-4   Scoring Criteria for Bearing Sleeve Parts

Assessment item		Assessment requirement	Assigned score	Scoring criteria	Test result	Score	Remark
1	Inner profile	ϕ45 mm/ Ra1.6 μm three places	40	2 points will be deducted for exceeding 0.01 mm and no point will be given for exceeding 0.02 mm			
		ϕ53 mm	5	No score for out-of-tolerance			
		ϕ30H7/ Ra1.6 μm	25	2 points will be deducted for exceeding 0.01 mm and no point will be given for exceeding 0.02 mm			
		ϕ32 mm×20	6	No score for out-of-tolerance			
		2×0.5	4	No score for out-of-tolerance			
		60±0.10 mm	6	No score for out-of-tolerance			
		8±0.05 mm	6	No score for out-of-tolerance			
		C2，C1 four places	2×4	No score for nonconformity			

Continue Table

Assessment item		Assessment requirement	Assigned score	Scoring criteria	Test result	Score	Remark
2	Process, program	Provisions on process and program		Deduct 1-5 points for violation of regulations			
3	Specification Operation	Rules related to standard operation of CNC lathe.		Deduct 1-5 points for violation of regulations			
4	Safe and civilized production	Relevant provisions on safe and civilized production		Deduct 1-50 points for violation of regulations			

Summary and feedback during students task implementation:

Teacher's review:

Item 4

Programming and Machining of

Complex Parts

[Item description]

Learning the content of this item requires mastering the preparation of typical shaft sleeve fit parts machining programs and the selection of machining tools; mastering the machining methods of fit parts, and the machining and inspection methods of shaft sleeve fit parts. Furthermore, theory should be contacted with practice, specifically carrying out simulation machining with NC simulation software in the programming classroom, entering the training workshop site for machine tool processing, and completing the machining tasks in groups or separately.

[Learning objectives]

（1）Master the preparation of typical shaft sleeve fit part machining programs.

（2）Master the machining methods of fit parts.

（3）Master the machining and inspection methods of shaft sleeve fit parts.

（4）Be able to comprehensively apply the learned professional theoretical knowledge to analyze the machining process of the complex parts.

（5）Be able to select the correct processing scheme and the appropriate tool according to the technical requirements of the parts, and be able to select the appropriate clamping method.

（6）Master the programming of typical arbor parts and the reasonable selection of cutting amount.

（7）Master the causes and prevention and elimination methods of errors in machining.

Task I　　Shaft sleeve fit parts machining

I. Drawings and technical requirements

Fig. 4-1-1 shows a set of shaft sleeve fit composite parts, the total fit length of the outer tapered surface within the shaft and sleeve is 78 ± 0.1 mm, and the total fit length of the outer cylindrical surface within the shaft and sleeve is 103 ± 0.1 mm. The material is 45 steel, blank: $\phi45$ mm $\times75$ mm, $\phi60$ mm $\times53$ mm, with one rod respectively.

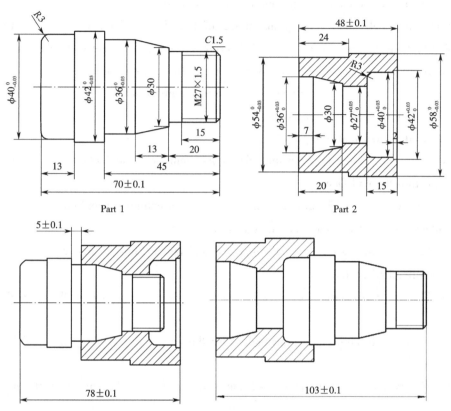

Part 1　　　　　　　　　　　　　　Part 2

Fig. 4-1-1　Fit Parts of Shaft Sleeve

Technical requirements.

(1) The sharp edge is deburred and chamfered, $C1$ is undeclared chamfer, and the center hole is allowed on the right end face of the shaft.

(2) Contact between tapered surfaces shall be more than 70%.

(3) Ra values of the shaft and sleeve are 1.6 μm.

II. Analysis of drawings and process arrangement

1. Analysis of drawings

Analysis of combination accuracy After the workpiece is combined, the dimension accuracy that is difficult to ensure mainly includes: fit length dimension (103 ± 0.1) mm, (78 ± 0.1) mm, clearance dimension (5 ± 0.1) mm. Other combination accuracy difficult to ensure is that the contact area of the inner and outer tapers is greater than 70%, the inner and outer R arcs and the inner and outer cylindrical surfaces fit. Analysis of machining method The difficulty in machining this part is to ensure the accuracy of each fit.

2. Process arrangement

1）Determine the clamping scheme of the workpiece

Both the rough and finishing machining of the shaft and sleeve need two times of clamping that can be completed. During NC turning, the starting point of machining is set at 2mm away from the workpiece blank. Axial cutting should be adopted as far as possible to improve the rigidity of workpiece and tool in the machining process.

2）Determine the machining route

（1）Part 1 shaft machining route.

① Use scroll chuck to align and clamp $\phi 45$ mm cylindrical blank 50mm out of the chuck and turn the end face.

② Rough turning $\phi 43$ mm，$\phi 36$ mm，$\phi 26.9$ mm cylindrical and outer taper with the radial allowance of 0.5mm and axial allowance of 0.1 mm.

③ Finishing turn $\phi 43$ mm，$\phi 36$ mm，$\phi 26.9$ mm cylindrical and outer taper to the dimensions required by the drawing.

④ Machine the external thread and check the thread accuracy by the go and no-go gauges.

⑤ Align the turning clamp，manually turn the end face，and keep the total length of 70 ± 0.1 mm.

⑥ Rough turn $\phi 40$ mm and $\phi 42$ mm cylindrical and $R3$ mm arc with the radial allowance of 0.5 mm and axial allowance of 0.1 mm.

⑦ Finishing turn $\phi 40$ mm and $\phi 42$ mm cylindrical and $R3$ mm arc to the dimensions required by the drawing.

⑧ Remove the workpiece and deburr and chamfer the workpiece.

（2）Part 2 sleeve machining route.

① Use scroll chuck to align and clamp $\phi 60$ mm cylindrical blank 30 mm out of the chuck and drill the hole manually with a diameter of $\phi 25$ mm. Turn the end face.

② Rough turn $\phi 36$ mm and $\phi 27$ mm inner arcs with 0.5 mm radial allowance.

③ Finishing turn $\phi 36$ mm and $\phi 27$ mm arcs to drawing size.

④ Turn to fit the inner and outer tapers and correct the inner taper surface to ensure the fit clearance is 5 ± 0.1 mm.

⑤ Rough turning $\phi 59$ mm and $\phi 54$ mm cylindrical with a radial allowance of 0.5 mm.

⑥ Finishing turn $\phi 54$m m cylindrical to the dimension required by the drawing.

⑦ Turn around and clamp at the place where the diameter is $\phi 54$ mm，correct $\phi 59$ mm cylindrical surface with a dial indicator，and finish $\phi 58$ mm cylindrical.

⑧ Manually turn the right end face to ensure 48 ± 0.1mm.

⑨ Rough turn ϕ42 mm and ϕ40 mm inner taper with radial allowance 0.5 mm for R3 mm arc.

⑩ Finishing turn ϕ42 mm and ϕ40 mm inner arcs and R3 mm arc to drawing size and ensure the fit length is（103±0.1）mm.

Remove the workpiece, deburr and chamfer, and check the machining accuracy.

3）Fill in the machining tool card and process card（see Table 4-1-1）

Table 4-1-1   Machining Tool Card and Process Card

Part drawing No.			4-1-1		CNC lathe process card	Machine model	CKA6150
Part name			Fit parts of shaft sleeve			Machine tool number	
Table of turning tools						Table of measuring tools	
S/N	Tool number	Tool compensation number	Name of turning tools	Turning tool parameters		Measuring tool name	Specification/ mm
1	T01	01	90° cylindrical rough turning tool	Type C blade		Vernier caliper Micrometer	25-50/0.01 50-75/0.01 0-150/0.02
2	T02	02	90° cylindrical finishing turning tool	Type D blade		Vernier caliper Micrometer	25-50/0.01 50-75/0.01 0-150/0.02
3	T03	03	External threading tool	Tool tip angle 60°		Go and no-go gauges	M27×1.5
4	T04	04	91° boring tool	Type C blade		Bore dial indicator	18-50/0.01
5			Bit ϕ25			Vernier caliper	0-150/0.02

Working procedure	Process content	Cutting amount			Machining Property
		S/（r/min）	F/（mm/r）	a_p/mm	
CNC lathe （shaft）	Use scroll chuck to clamp blank to turn the end face.	600-800	0.2	—	Manual
1	The outer profile of right end of rough and finish turning shaft.	600-1 200	0.1-0.2	0.25-2	Automatic
2	Machine the external thread.	600	1.5	—	Automatic
CNC lathe	Align the turning clamp, manually turn the end face, and keep the total length of 70±0.1 mm.	600-800	0.2	—	Manual
1	The outer profile of left end of rough and finish turning shaft.	600-1 200	0.1-0.2	0.25-2	Automatic

2	Remove the workpiece and deburr and chamfer the workpiece.	—	—	—	—
CNC lathe (sleeve)	Manually drill through holes with scroll chuck, with a diameter of ϕ 25 mm. Radial turning.	300-400	—	—	Manual
1	The inner profile of left end of rough and finish turning sleeve.	600-1 200	0.1-0.2	0.25-2	Automatic
2	Match inner and outer tapers.	1 200	0.1	—	Automatic
3	The outer profile of left end of rough and finish turning sleeve.	600-1 200	0.1-0.2	0.25-2	Automatic
CNC lathe	Turn around and clamp ϕ54 mm cylindrical, and correct ϕ59 mm cylindrical surfaces with a dial indicator.				
1	Turn end face to ensure 48±0.1mm.	600-800	0.2	—	Manual
2	The inner and outer profiles of right end of rough and finish turning sleeve.	600-1 200	0.1-0.2	0.25-2	Automatic
3	Deburr and chamfer, and check the machining accuracy.				

III. Program Preparation and Part Machining

Due to the low requirements for the length of the workpiece, according to the principle of determining the program origin, the program origin of the workpiece can be taken at the intersection of the right end face of the workpiece and the spindle axis after clamping.

According to the part shown in Fig. 4-1-1, the machining route of the workpiece is analyzed, and the clamping scheme during machining, together with the tool and cutting amount, is determined. It is divided into four parts according to the machining procedures, and four corresponding programs are prepared to complete the machining. Table 4-1-2 shows the program for machining the external profile of the right end of the shaft, Table 4-1-3 shows the program for machining the external profile of the left end of the shaft, Table 4-1-4 shows the program for machining the internal profile of the left end of the sleeve, and Table 4-1-5 shows the program for machining the internal profile of the right end of the sleeve.

Table 4-1-2 Program for Machining the External Profile of the Right End of the Shaft

Program content	Program description
04101;	Program number;
N1;	1st program segment number;
G99 M03 S600 T0101;	Select #1 tool, spindle forward rotation, 600 r/min;
G00 X100.0 Z100.0;	Rapid movement to safe point;
G00 G42 X45.0 Z2.0;	Rapidly move to the loop point and add tool tip arc radius right compensation;
	Coolant on;
M08;	Rough machining loop;
G71 U1.0 R0.5;	
G71 P10 Q20 U0.5 W0.1 F0.2;	
N10 G00 X24.0;	
G01 Z0. F0.1;	Starting program segment of loop machining;
G01 X26.9. Z-1.5;	
G01 Z-20.;	
G01 X30. ;	
G01 X36.W-13.;	
G01 Z-45.;	
G01 X43.;	
G01 W-5.;	
N20 G00 X45.;	
G00 X100.0 Z100.0;	End program segment of loop machining;
M05;	Rapid movement to safe point;
M00;	Coolant off;
N2;	Program suspension;
G99 M03 S1200 T0202;	2nd program segment number;
G00 X100.0 Z100.0;	Select #2 tool, spindle forward rotation, 1200 r/min;
G00 X45.0 Z2.0;	Rapid movement to safe point;
G70 P10 Q20;	Rapid movement to loop point;
G00 G40 X100.0 Z100.0;	Finishing loop;
M05;	Rapidly move to safe point and cancel tool compensation;
M00;	Spindle stop;
N3;	Program suspension;
G99 M03 S600 T0303;	3rd program segment number;
G00 X28. Z5.;	Select #3 tool, spindle forward rotation, 600 r/min;
G92 X26.4 Z-15. F1.5;	Rapidly move to the starting point of threat cutting loop;
X25.9;	Thread cutting loop;
X25.5;	
X25.2;	
X25.1;	
X25.05;	
G00 X100.Z100.;	
M30;	Retract tool;
	End of program and return to the 1st program segment

Table 4-1-3　Program for Machining the External Profile of the Left End of the Shaft

Program content	Program description
04102;	Program number;
N1;	1st program segment;
G99 M03 S600 T0101;	Select #1 tool, spindle forward rotation, 600 r/min;
G00 X100.0 Z100.0;	Rapid movement to safe point;
G00 X46.0 Z5.0;	Rapid movement to loop point;
M08;	Coolant on;
G71 U1.0 R0.5;	Rough machining loop;
G71 P10 Q20 U0.5 W0.1 F0.2;	
N10 G00 X34.0;	Starting program segment of loop machining;
G01 Z0.F0.1;	
G03 X40 W-3.R3.;	
G01 Z-13.;	
G01 X42.0;	
G01 W-13.;	
N20 G00 X46.;	End program segment of loop machining;
G00 X100. Z100.;	Rapid movement to safe point;
M09;	Coolant off;
M05;	Spindle stop;
M00;	Program suspension;
N2;	2nd program segment number;
G99 M03 S1200 T0202;	Select #2 tool, spindle forward rotation, 1200 r/min;
G00 G42 X46.0 Z5.0;	Rapidly move to the loop point, tool right compensation;
M08;	Coolant on;
G70 P10 Q20;	Finishing loop;
G00 G40 X100. Z100.;	Rapidly move to safe point and cancel tool tip arc radius compensation;
M09;	Coolant off;
M30;	End of program and return to the 1st program segment

Table 4-1-4　Program for Machining the Internal Profile of the Left End of the Sleeve

Program content	Program description
04103;	Program number;
N1;	1st program segment number;
G99 M03 S600 T0404;	Select #4 tool, spindle forward rotation, 600 r/min;
G00 X100.0 Z100.0;	Rapid movement to safe point;
G00 G41 X25.0 Z2.0;	Rapidly move to the loop point, tool left compensation;
M08;	Coolant on;
G71 U1.0 R0.5;	Rough machining loop;

Continue Table

Program content	Program description
G71 P10 Q20 U−0.5 W0.05 F0.2;	
N10 G00 X36.0;	Starting program segment of loop machining;
G01 Z−7.F0.1;	
G01 X30.W−13.;	
G01 X27.;	
G01 W−15.;	
N20 G00 X25.;	End program segment of loop machining;
G00 Z100.;	Rapid movement to safe point;
G00 X100.;	
M09;	Coolant off;
M00;	Program suspension;
N2;	2nd program segment number;
G99 M03 S1200 T0404;	Select #4 tool, spindle forward rotation, 1200 r/min;
G00 X25.0 Z2.0;	Rapid movement to loop point;
M08;	Coolant on;
G70 P10 Q20;	Finishing loop (skip taper);
G00 G40 X100.Z100.;	Rapid movement to safe point;
M09;	Coolant off;
M30;	End of program and return to the 1st program segment

Table 4-1-5　Program for Machining the Internal Profile of the Right End of the Sleeve

Program content	Program description
04104;	Program number;
N1;	1st program segment number;
G99 M03 S600 T0404;	Select #4 tool, spindle forward rotation, 600 r/min;
G00 X100.0 Z100.0;	Rapid movement to safe point;
G00 G41 X25.0 Z2.0;	Rapidly move to the loop point, tool left compensation;
M08;	Coolant on;
G71 U1.0 R0.5;	Rough machining loop;
G71 P10 Q20 U−0.5 W0.F0.2;	
N10 G00 X42.;	Starting program segment of loop machining;
G01 Z−2.;	
G01 X40.;	
G01 Z−12.	
G03 X34.W−3.R3.	
N20 G01 X25.;	End program segment of loop machining;
G00 Z100.;	
G00 X100.	Rapid movement to safe point;
M09;	Coolant off;
M00;	Program suspension;

Continue Table

Program content	Program description
N2;	2nd program segment number;
G99 M03 S1200 T0404;	Select #4 tool, spindle forward rotation, 1200 r/min;
G00 X25.0 Z2.0;	Rapid movement to loop point;
M08;	Coolant on;
G70 P10 Q20;	Finishing loop;
G00 Z100.;	Rapid movement to safe point;
G00 G40 X100.	Cancel tool compensation;
M09;	Coolant off;
M30;	Return to program start after program end

IV. Scoring Table

Task evaluation as per Table 4-1-6.

Table 4-1-6 Scoring Criteria for Shaft Sleeve Fit Parts

	Assessment item		Assessment requirement	Assigned score	Scoring criteria	Test result	Score	Remark
1	Part 1	Outer diameter	$\phi 42_{-0.03}^{0}$ mm / $Ra1.6$ μm	4+1	Full deduction for out-of-tolerance			
2			$\phi 40_{-0.03}^{0}$ mm/ $Ra1.6$ μm	4+1	Full deduction for out-of-tolerance			
3			$\phi 36_{-0.03}^{0}$ mm/ $Ra1.6$ μm	4+1	Full deduction for out-of-tolerance			
4	Part 1	Length	20 mm	2	Full deduction for out-of-tolerance			
5			45 mm	2	Full deduction for out-of-tolerance			
6			13 mm (2 places)	4	Full deduction for out-of-tolerance			
7			70±0.1 mm	3	Full deduction for out-of-tolerance			
8		Arc	$R3$ mm/ $Ra1.6$ μm	2+1	R gauge			
9		Thread	M27×1.5/ $Ra3.2$ μm	6+2	Thread go and no-go gauges measurement			
10		Chamfer	C1.5	2	Visual inspection			

Continue Table

Assessment item			Assessment requirement	Assigned score	Scoring criteria	Test result	Score	Remark
11	Part 2	Outer diameter	$\phi 54^{\ 0}_{-0.03}$ mm/ Ra1.6 μm	4+1	Deduct 2 points for out-of-tolerance 0.01 mm			
12			$\phi 58^{\ 0}_{-0.03}$ mm/ Ra1.6 μm	4+1	Deduct 2 points for out-of-tolerance 0.01 mm			
13		Inner diameter	$42\phi^{+0.03}_{\ 0}$ mm/ Ra1.6 μm	4+1	Deduct 2 points for out-of-tolerance 0.01 mm			
14			$40\phi^{+0.03}_{\ 0}$ mm/ Ra1.6 μm	4+1	Deduct 2 points for out-of-tolerance 0.01 mm			
15			$\phi 36^{+0.03}_{\ 0}$ mm/ Ra1.6 μm	4+1	Deduct 2 points for out-of-tolerance 0.01 mm			
16			$\phi 27^{+0.03}_{\ 0}$ mm/ Ra1.6 μm	4+1	Deduct 2 points for out-of-tolerance 0.01 mm			
17	Part 2	Length	2 mm	2	Full deduction for out-of-tolerance 0.2 mm			
18			7 mm	2	Full deduction for out-of-tolerance 0.2 mm			
19			13 mm	2	Full deduction for out-of-tolerance 0.2 mm			
20			15 mm	2	Full deduction for out-of-tolerance 0.2 mm			
21			24 mm	2	Full deduction for out-of-tolerance 0.2 mm			
22			48±0.1 mm	2	Full deduction for out-of-tolerance 0.2 mm			
23		Arc	R3 mm/ Ra1.6 μm	2+1	Visual inspection			
24	Fit clearance		5±0.1 mm	4	Full deduction for out-of-tolerance			
25	Fit length		103±0.1 mm	4	Full deduction for out-of-tolerance			
			78±0.1 mm	4	Full deduction for out-of-tolerance			
26	Taper fit		More than 60% of the contact surface	4	Visual inspection			

	Assessment item	Assessment requirement	Assigned score	Scoring criteria	Test result	Score	Remark
27	Process, program	Provisions on process and program		Deduct 1-5 points for violation of regulations			
28	Standardized operation	Rules related to standard operation of CNC lathe.		Deduct 1-5 points for violation of regulations			
29	Safe and civilized production	Rules related to safe and civilized production.		Deduct 1-5 points for violation of regulations			

Summary and feedback during students task implementation:

Teacher's review:

Task II Mold arbor part machining

I. Drawings and technical requirements

According to the arbor shown in Fig. 4-2-1, develop the machining scheme and complete the machining on the CNC lathe. Blank: $\phi50$ mm×125 mm bar. Material: 45 steel.

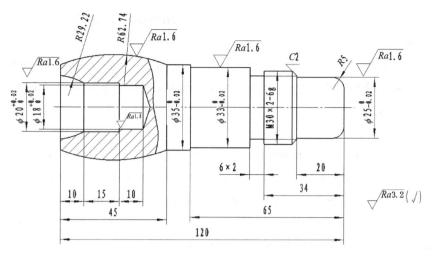

Fig. 4-2-1   Arbor Part

II. Analysis of drawings

1. Analysis of drawings

As shown in Fig. 4-2-1, the profile of the part is clearly described and the dimensions are completely marked. In terms of profile, the part has steps, grooves, threads and arcs on the external profile, and internal platforms and inner arcs on the internal profile,

1) Concept of arbor

Arbor that is used to support rotating parts only bears bending moment without transmitting torque. Some arbors rotate, such as the axles of railway vehicles, and some do not rotate, such as the axles supporting the pulley. As shown in Fig. 4-2-2

Fig. 4-2-2   Arbor

2) Classification of arbors

According to whether the arbor rotates during operation, it can be divided into rotating arbor and fixed arbor.

(1) Rotating arbor: Arbor bears bending moment and rotates during working.

(2) Fixed arbor: Arbor bears bending moment and is fixed during working.

3) Arbor function

According to the function of the shaft and the load borne, it can be divided into arbor, transmission shaft and drive shaft. The arbor only bears bending moment but does not transmit torque; the rotating shaft not only bears bending moment but also transmits torque; the transmission shaft only transmits torque but does not bear bending moment, or bears very low bending moment.

2. Process arrangement

1) Determine the clamping scheme of the workpiece

The arbor part has a total length of 120 mm. At its left end is an $R52.74$ mm arc surface with an internal profile being 35 mm long and an external profile being 45 mm long, and at its right end is a stepped shaft. Therefore, considering the overall shape of this part, clamping can be performed in the way of clamping one position and pressing against one position during the machining of $\phi25$ mm, $\phi30$ mm, $\phi33$ mm and $\phi35$ mm steps at the right end (a $\phi46$ mm$\times10$ mm process step can be built at the left end of the workpiece). The three-jaw chuck can be used to machine the workpiece to be clamped at the right end.

2) Determine the machining route

More surfaces shall be machined in one-time clamping from inside to outside, from coarse to fine, and from near to far to the maximum degree. Besides this, the surfaces of the inner hole can be machined when internal and external profiles are machined. The detailed machining sequence is shown below:

(1) Flatten the right end face, drill the center hole, and finish the cylindrical surface of the blank with its length of 75 mm.

(2) Turn around and perform alignment; use a three-jaw chuck to clamp the finished cylindrical surface; perform turning on the $\phi47$ mm cylindrical surface and the 10 mm-long process step; build process steps with a total length of 120.5mm.

(3) The method of clamping one position and pressing against one position is used to roughly turn and machine $\phi35$ mm, $\phi33$ mm, $\phi30$ mm, and $\phi25$ mm cylindrical surfaces, and R5 arc surface with a radial allowance of 0.5 mm, an axial allowance of 0.1 mm, and an axial length of 75 mm.

(4) Machine the 6 mm$\times2$ mm thread tool retraction groove with a grooving tool.

(5) Finely turn the $\phi35$ mm, $\phi33$ mm, $\phi30$ mm, and $\phi25$ mm cylindrical surfaces, and R5 arc surfaces to the drawing requirements.

(6) Use a threading tool to turn the M30$\times2$-6g external thread to the accuracy requirements.

(7) Turn around and use a three-jaw chuck for clamping; cushion a copper sheet to align the $\phi33$ mm cylindrical surface and firmly clamp it. Flatten the end face with a total length of 120 mm.

（8）Drill the B2 center hole.

（9）Use the ϕ16 bit to drill holes with the inner hole length of 35 mm.

（10）Roughly bore the internal profile including an R29.22 mm inner arc surface, a ϕ20 mm inner hole, and ϕ18 mm inner hole）.

（11）Roughly bore parts above till their dimensions comply with the drawing requirements.

（12）Roughly and finely turn the R62.74 mm outer arc surface to the drawing requirements.

3）Fill in the machining tool card and process card（see Table 4-2-1）

Table 4-2-1　Machining Tool Card and Process Card

Part drawing No.	4-2-1	CNC lathe process card	Machine model	CKA6150
Part name	Mold arbor part		Machine tool number	
Table of turning tools			Table of measuring tools	
Tool number	Tool compensation number	Name of turning tools	Turning tool parameters	Measuring tool name	Specification/mm
T01	01	90° cylindrical surface rough and fine turning tools	Type C blade	Vernier caliper Micrometer	0-150/0.02 25-50/0.01
T02	02	93° cylindrical surface rough and fine turning tools	Type D blade	Vernier caliper Micrometer	0-150/0.02 25-50/0.01
T03	03	Grooving tool	4.0 mm wide blade	Vernier caliper	0-150/0.02
T04	04	External threading tool	Tool tip angle 60°	Ring gauge	M30×2
T05	05	91° boring tool	Type D blade	Bore dial indicator	18-35/0.01
		Center drill	B2		
		Bit ϕ16 mm		Vernier caliper	0-150/0.02

Continue Table

Working procedure		Process content	Cutting amount			Processing property
			$S/$ (r/min)	$F/$ (mm/r)	a_p/mm	
CNC lathe	1	Use a three-jaw chuck for clamping, and turn the cylindrical surface and end face to determine the reference.	600	—	1	Manual
	2	Roughly machine ϕ35 mm, ϕ33 mm, ϕ30 mm, and ϕ25 mm cylindrical surfaces and R5 mm arc surface	500-700	0.2	2	Automatic
	3	Cut a 6×2 mm tool retraction groove.	400-600	0.05		Automatic
	4	Finely machine ϕ35 mm, ϕ33 mm, ϕ30 mm, and ϕ25 mm cylindrical surfaces and R5 mm arc surface	1 000-1 200	0.1	0.25	Automatic
CNC lathe	5	Machine the M30×2 external thread.	600	2.0	—	Automatic
CNC lathe	1	Turn around, and perform clamping and alignment to ensure that the total length of 120 mm.	500	0.1	—	Manual
	2	Manually drill the center drill.	800	—	—	Manual
	3	Drill the ϕ16 mm hole.	300	—	—	Manual
	3	Roughly bore the internal profile including the R29.22 inner arc surface, ϕ20 mm inner hole, and ϕ18 mm inner hole).	500-700	0.15	1.5	Automatic
	4	Finely bore the internal profile including the R29.22 mm inner arc surface, ϕ20 mm inner hole, and ϕ18 mm inner hole).	700-1 000	0.08	0.25	Automatic
	5	Roughly turn the R62.74 mm outer arc surface.	600-700	0.2	1	Automatic
	6	Finely turn the R62.74 mm outer arc surface to the drawing requirements.	1 000-1 200	0.1	0.25	Automatic

III. Program Preparation and Part Machining

According to the part shown in Fig. 4-2-1, the machining route of the workpiece is analyzed, and the clamping scheme during machining, together with the tool and cutting amount, is determined. It is divided into 3 parts according to the machining content, and 3 corresponding programs are prepared to complete the machining. Table 4-2-2 shows the program for machining the external profile at the right end of the part; Table 4-2-3 shows the program for machining the internal profile at the left end of the part; Table 4-2-4 shows the program for machining the external profile at the left end of the part.

Table 4-2-2 Program for Machining the External Profile at the Right End of the Part

Program content	Program description
O4201;	Program number;
N1;	1st program segment;
G99 M03 S600 T0101;	Select #1 tool, spindle forward rotation, 600 r/min;
G00 X200.0 Z10.0;	Rapid movement to safe point;
G00 X50.0 Z2.0;	Rapid movement to loop point;
M08;	Coolant on;
G71 U2.0 R0.5;	Rough machining loop;
G71 P10 Q20 U0.5 W0.05 F0.2;	
N10 G00 X15.0;	Starting program segment of loop machining;
G01 Z0.0 F0.1;	
G02 X25. Z-5.0 R5.0;	
G01 Z-20.0;	
G01 X26.0	
G01 X29.9 W-2.0;	
G01 Z-40.0;	
G01 X33.0;	
G01 Z-65.0;	
G01 X35.0;	
G01 W-10.0;	
N20 G01 X50.0;	End program segment of loop machining;
G00 X200.0;	Retract tool;
G00 Z10.0;	
M09;	Coolant off;
M05;	Spindle stop;
M00;	Program suspension;
N2;	2nd program segment;
G99 M03 S500 T0303;	Select #3 tool (grooving tool); spindle forward rotation, 500 r/min;
G00 X35.0 Z-40.0;	Rapid movement to tool starting point;
M08;	Coolant on;
G01 X26.0 F0.05;	Grooving;
G00 X35.0;	Retract tool;

Program content	Program description
G00 W2.;	Axial movement by 2 mm;
G01 X26.0 F0.05;	Grooving;
G00 X200.0;	Radial retraction;
G00 Z10.;	Axial retraction;
M09;	Coolant off;
M05;	Spindle stop;
M00;	Program suspension;
N3;	3rd program segment;
G99 M03 S1200 T0101;	Select #1 tool (90° cylindrical forward offset tool); spindle forward rotation, 1200 r/min;
G00 G42 X50.0 Z2.0;	Adjust the tool to the starting point of loop and add tooltip arc right compensation;
M08;	Coolant on;
G70 P10 Q20;	Finishing loop;
G00 G40 X200.0 Z10.0;	Rapidly move to safe point and cancel tool tip arc radius compensation;
M09;	Coolant off;
M05;	Spindle stop;
M00;	Program suspension;
N4;	4th program segment;
G99 M03 S600 T0404;	Select #4 tool (external threading tool); spindle forward rotation, 600 r/min;
G00 X35.0 Z-15.0;	Rapid movement to tool starting point of loop;
M08;	Coolant on;
G92 X29.2 Z-37.0 F2.0;	Thread cutting loop;
X28.6;	
X28.1;	
X27.7;	
X27.5;	
X27.4;	
G00 X200.0 Z10.0;	Retract tool;
M09;	Coolant off;
M05;	Spindle stop;
M30;	End of program and return to the 1st program segment

Table 4-2-3 Program for Machining the Internal Profile at the Left End of the Part

Program content	Program description
04202;	Program number;
N1;	1st program segment;
G99 M03 S600 T0505;	Select #5 tool; spindle forward rotation, 600 r/min;
G00 X100.0 Z100.0;	Rapid movement to safe point;
G00 X16.0 Z2.0;	Rapid movement to loop point;
M08;	Coolant on;
G71 U1.5 R0.5;	Rough machining loop;
G71 P10 Q20 U-0.5 W0.05 F0.15;	
N10 G00 X26.06;	Starting program segment of machining;
G01 Z0.0 F0.1;	
G03 X20.0 Z-10.0 R29.22;	
G01 W-15.0;	
G01 X18.0;	
G01 W-10.0;	
N20 G01 X16.0;	End program segment of loop machining;
G00 Z300.0;	Retract tool;
G00 X100.0;	
M09;	Coolant off;
M05;	Spindle stop;
M00;	Program suspension;
N2;	2nd program segment;
G99 M03 S1200 T0505;	Select #5 tool; spindle forward rotation, 1200 r/min;
G00 X100.0 Z100.0;	Rapid movement to safe point;
G00 G41 X16.0 Z2.0;	Rapid movement to loop point and adding tooltip arc radius left compensation;
M08;	Coolant on;
G70 P10 Q20;	Finishing loop;
G00 Z300.0;	Retract tool;
G00 G40 X200.0;	Cancel tool compensation;
M09;	Coolant off;
M05;	Spindle stop;
M30;	End of program and return to the 1st program segment

Table 4-2-4　Program for Machining the External Profile at the Left End of the Part

Program content	Program description
04203;	Program number;
N1;	1st program segment;
G99 M03 S600 T0202;	#2 tool selected，spindle forward rotation，600 rpm;
G00 X100.0 Z100.0;	Rapid movement to safe point;
G00 X50.0 Z2.0;	Rapid movement to loop point;
M08;	Coolant on;
G71 U1.0 R0.5;	Rough machining loop;
G71 P10 Q20 U0.5 W0.05 F0.2;	
N10 G00 X35.0;	Starting program segment of loop machining;
G01 Z0. F0.1;	
G03 X35.0 Z−45.0 R52.74;	
N20 G01 X50.0;	End program segment of loop machining;
G00 X100.0 Z100.0;	Rapid movement to safe coolant off;
M05;	Spindle stop;
M09;	Coolant off;
M00;	Program suspension;
N2;	2nd program segment;
G99 M03 S1200 T0202;	Select #2 tool，spindle forward rotation，1200 r/min;
G00 G42 X50.0 Z2.0;	Rapidly move to the loop point，tool right compensation;
M08;	Coolant on;
G70 P10 Q20;	Finishing loop;
G00　Z300.0;	Rapid movement to safe point;
G00　G40 X200.0;	Cancel tool compensation;
M09;	Coolant off;
M05;	Spindle stop;
M30;	End of program and return to the 1st program segment

IV. Task assessment

Task evaluation as per Table 4-2-5.

Table 4-2-5　Scoring Criteria for Arbor Part

Assessment item		Assessment requirement	Assigned score	Scoring criteria	Test result	Score	Remark
1	Outer diameter	$\phi 42_{-0.02}^{0}$ mm	8	Deduct 4 points for out-of-tolerance 0.01 mm			
2		$\phi 33_{-0.02}^{0}$ mm	8	Deduct 4 points for out-of-tolerance 0.01 mm			
3		$\phi 35_{-0.02}^{0}$ mm	8	Deduct 4 points for out-of-tolerance 0.01 mm			
4	Inner diameter	$\phi 20_{0}^{+0.02}$ mm	8	Deduct 4 points for out-of-tolerance 0.01 mm			
5		$\phi 18_{0}^{+0.02}$ mm	8	Deduct 4 points for out-of-tolerance 0.01 mm			
6	Arc	$R5$ mm	8	Detection with R gauge			
7		$R62.74$ mm	8	Detection with R gauge			
8		$R29.22$ mm	8	Detection with R gauge			
9	Thread	M30×2-6g	14	Thread go/no-go gauge			
10	Length	8 places	8	Full deduction for out-of-tolerance			
11	Surface	Ra1.6 μm at 8 places	8	No score for 1 level higher for Ra			
12	Tool retraction groove	6 mm×2 mm	6	Full deduction for out-of-tolerance			
13	Process, program	Provisions on process and program	Deduct 1-5 points for violation of regulations				
14	Standardized operation	Rules related to standard operation of CNC lathe.	Deduct 1-5 points for violation of regulations				
15	Safe and civilized production	Rules related to safe and civilized production.	Deduct 1-5 points for violation of regulations				

Continue Table

Assessment item	Assessment requirement	Assigned score	Scoring criteria	Test result	Score	Remark
Summary and feedback during students task implementation:						
Teacher's review:						